How Fast Can a Falcon Dive?

Animal Q&A: Fascinating Answers to Questions about Animals

Animal Q&A books invite readers to explore the secret lives of animals. Covering everything from their basic biology to their complex behaviors at every stage of life to issues in conservation, these richly illustrated books provide detailed information in an accessible style that brings to life the science and natural history of a variety of species.

Do Butterflies Bite?: Fascinating Answers to Questions about Butterflies and Moths, by Hazel Davies and Carol A. Butler

Do Bats Drink Blood? Fascinating Answers to Questions about Bats, by Barbara A. Schmidt-French and Carol A. Butler

Why do Bees Buzz? Fascinating Answers to Questions about Bees, by Elizabeth Capaldi Evans and Carol A. Butler

Do Hummingbirds Hum? Fascinating Answers to Questions about Hummingbirds, by George C. West and Carol A. Butler

How Fast Can a Falcon Dive?

Fascinating Answers to Questions about Birds of Prey

Peter Capainolo
and Carol A. Butler

Rutgers University Press

NEW BRUNSWICK, NEW JERSEY, AND LONDON

Library of Congress Cataloging-in-Publication Data

Capainolo, Peter, 1959–
  How fast can a falcon dive? : fascinating answers to questions about
birds of prey / Peter Capainolo and Carol A. Butler.
    p. cm. — (Animal Q & A)
  Includes bibliographical references and index.
  ISBN 978–0–8135–4790–9 (pbk. : alk. paper)
  1. Birds of prey—Miscellanea. I. Butler, Carol A. II. Title.
  QL677.78C37 2010
  598.9—dc22

                                                              2009048292

A British Cataloging-in-Publication record for this book is available
from the British Library.

Illustrations on pages xiv–xvii copyright Windsor Nature Discovery/
Karen Pidgeon. Used with permission. www.nature-discovery.com

Visit our Web site: http://rutgerspress.rutgers.edu

Manufactured in the United States of America

Contents

Preface and Acknowledgments   ix

ONE Raptor Basics   1

1 What is a raptor?   1
2 Where in the world are raptors found?   2
3 What is the connection between "raptor" dinosaurs and modern birds of prey?   5
4 How are raptors classified?   8
5 What are the differences between falcons and hawks?   14
6 Are all owls nocturnal?   16
7 Are eagles the largest raptors?   18
8 Are vultures the only raptors that eat dead animals?   18
9 What is a buzzard?   19
10 Which raptor is the smallest?   20
11 How long do raptors live in the wild?   20

TWO Raptor Bodies   22

1 How do male and female raptors differ?   22
2 What do birds of prey eat?   23
3 How much does a raptor eat in a day?   25
4 How do raptors digest their food?   28
5 Can the species of a raptor be identified by its waste pellets?   29
6 Are raptors warm- or cold-blooded?   31

7 What are the features of raptors' wings? 32

8 Do raptors molt? 35

9 What is "preening"? 36

10 What is special about raptors' feet? 37

11 How well do birds of prey hear? 38

12 How well can raptors see? 39

13 Do raptors have a keen sense of smell? 44

THREE Raptor Behavior 46

1 How intelligent are birds of prey? 46

2 How fast can a raptor fly? 47

3 How fast can a falcon dive? 50

4 How far can a raptor fly? 53

5 Do all raptors migrate? 55

6 How do raptors find their way during migration? 56

7 How do raptors hunt? 59

8 Are birds of prey social or loners? 60

9 How do raptors communicate? 61

10 Are raptors always aggressive? 61

11 Are raptors dangerous to people? 63

FOUR Raptor Reproduction 65

1 How do birds of prey attract a mate? 65

2 At what time of year do birds of prey mate? 65

3 How do raptors mate? 68

4 Are raptors monogamous? 71

5 Do raptors of one species mate with other species? 74

6 How is artificial insemination practiced with raptors? 75

7 Do all birds of prey make nests? 78

8 Do raptor parents share nesting duties? 83

9 How many eggs do various species of raptors lay? 84

10 What do raptor eggs look like? 85

11 What is the likelihood of survival for a raptor
    nestling? 86

12 How does a raptor female feed her nestlings? 86

13 How long does it take before raptor young fledge? 87

FIVE Dangers and Defenses   89

1 Which animals prey on raptors?   89
2 How do prey animals defend themselves against
   raptors?   89
3 How do raptors defend themselves?   91
4 What illnesses occur in wild raptors?   92
5 What injuries are common among wild raptors?   93
6 What other dangers do raptors face as a result of
   development and population growth?   101
7 Does lead in the environment affect raptors?   103
8 Has DDT affected birds of prey?   104
9 Which raptors are particularly vulnerable to environmental
   toxins?   106
10 Do other environmental toxins endanger birds of
   prey?   110

SIX Raptor Husbandry   113

1 What is meant by "husbandry"?   113
2 How do zoos and rehabilitation facilities house
   raptors?   113
3 What does a rehabilitator do with a sick or injured
   raptor?   114
4 What is "imprinting"?   116
5 What are raptors fed in captivity?   118
6 How are raptors housed for falconry?   119
7 How long do raptors live in captivity?   120

SEVEN Taming and Training   121

1 What is "falconry"?   121
2 Where and when did humans begin using captive raptors
   for hunting?   123
3 What role do dogs play in falconry?   131
4 How and where can you acquire a raptor to train?   132
5 Do you always need a license to possess a raptor?   136
6 Can all raptors be tamed and trained?   138

7 How are raptors trained? 140
8 Can a trained raptor be released back into the wild? 143
9 What is the status of falconry today? 144

EIGHT Raptors and People 147

1 Have attitudes about raptors changed over time? 147
2 Why are people fascinated by raptors? 149
3 Where can I see raptors? 155
4 What should I do if I find an injured or dead raptor? 156
5 What attracts raptors to live in cities? 159
6 Of what value are raptors to the environment? 162

NINE Research and Conservation 163

1 Why do we need to study raptors? 163
2 How do museums accumulate their collections of
   specimens? 168
3 How are raptors captured for study? 169
4 How are raptor skins prepared for study or exhibit? 170
5 Are any raptors endangered? 171

Appendices 173

A Places to See Birds of Prey 173
B Recommended Reading and Web Sites 177
C Raptor Species Mentioned in the Book 179

References 189
Index 209

A color insert follows page 124

Preface and Acknowledgments

Birds of prey have stirred human emotions since time immemorial because of their fierce beauty, great strength, commanding presence, and superlative skill as hunters. . . . Today, birds of prey are arguably the most popular group of birds . . . attracting the most ardent and zealous devotees, whether they be falconers, rehabilitators, breeders, banders, scientists, or birdwatchers.

—Tom J. Cade, foreword,
Understanding the Bird of Prey by Nick Fox

The human fascination with raptors extends far into the past. As we can see in Medieval artwork, knights were so involved in taming and displaying their prized hawks that they took them along everywhere "on the glove," even to church. When you hunt with a falcon, according to falconer Stephen Bodio, "your purpose in the field is to assist the bird, your reward the companionship of a creature that could disappear over the horizon in fifteen seconds flat." Avid falconer Tom Cade puts it this way: "To have such a magnificent creature as a hawk, falcon, or eagle accept you as its companion and to allow you to enter its space is to gain a rare perspective on life." Bil Gilbert describes his own obsession with falconry as so exhausting and all consuming that he finally gave it up and limited himself to writing about birds and banding them. Alvah Nye, one of the first people to train hawks in the United States, reduces the falconer's role in the field even further: "You are the bird."

The colorful history of falconry reflects just one aspect of our relationship with birds of prey. Birds were likely to be objects of interest mainly because they provided a good meal until eighteenth-century botanist Carl Linnaeus and his contemporaries began to study and categorize the natural world. John James Audubon and other artists began publishing collections of naturalistic illustrations of birds that they often drew from birds they had killed on expeditions for that purpose. These drawings had scientific as well as artistic value, and this scientific focus contributed to the nineteenth-century mania for collecting specimens of birds and their eggs. When the conservation movement took hold in the 1930s, conservationists actively discouraged wanton destruction of wildlife and created awareness that some species were in danger of becoming extinct. For some, this social pressure stigmatized hunting as a hobby, and collecting and hunting became much less popular. The publication of field identification guides and the development of moderately priced binoculars and cameras led to the present-day interest in observing raptors and other birds in conjunction with efforts to preserve their numbers. The International Association for Falconry and Conservation of Birds of Prey (IAF) currently represents seventy associations from forty-eight countries that have a combined membership of over thirty-five thousand.

In this book, we explore the biology and behavior of birds of prey, their evolution, and their place in history. Which birds are classified as birds of prey? What physical characteristics equip them to pursue their prey? How do they mate and raise their young? Are they endangered? What restrictions affect their handling and training? Coauthor Peter Capainolo shares some of his personal experiences with raptors from the point of view of a falconer, as well as some of the details of his work with the collections in the Ornithology Department of the American Museum of Natural History in New York. His colleagues at the museum have been generous with their contributions and support. The book is illustrated with beautiful photographs by the renowned nature photographer Richard Ettlinger, as well as with photographs and drawings contributed by people from the global community of bird enthusiasts who were kind enough to support this project.

Our source for spelling and punctuation of common and scientific names is the third edition of *The Howard and Moore Complete Checklist of the Birds of the World*, edited by E. C. Dickinson (Princeton University Press). When we use the common name of a raptor, we follow it with the scientific name only on the first occasion it is mentioned in each chapter.

Once we have mentioned a genus in a chapter, e.g., *Falco punctatus*, we describe the next bird mentioned in that genus using only the first initial of the genus, e.g., *F. peregrinus*. Following conventional usage, common names of birds are capitalized, and common names of mammals are lowercased.

Appendix A provides information about where to watch raptors in the wild, and appendix B offers some print and online resources for readers who want additional information. A species list in appendix C lists all the raptors mentioned in the book along with their scientific names and where they can be found. The reference section lists sources for some of the information in our answers.

If you have ever been fascinated by these fierce and wild birds and want to know more about them, we offer this book as your resource.

The production of this book was a group endeavor, especially because the subject of birds of prey is so diverse. Dan Jacobs introduced us to one another, never suspecting his serendipitous action would spawn a book, and we thank him—we have enjoyed working together. Richard Ettlinger's superlative photos excited us early on when we saw some of them on exhibit at the American Museum of Natural History (AMNH), and we cannot thank him enough for e-mailing us two big files of photos and giving us our choice for the book. Due to printing constraints, we could select only a limited number, a harrowing task for us since every one of his shots would have enhanced our book. We are grateful to ornithologists Jeff Groth and George West for their advice on the manuscript and for tirelessly working on illustrations and photos whenever we needed their superb computer graphics skills. At the Department of Ornithology, AMNH, Joel Cracraft, curator-in-charge, and Paul Sweet, collections manager, were very supportive and generous to us in various ways and we thank them for their kindness. We appreciate the inspiration and encouragement that was always forthcoming from our editor, Doreen Valentine, the folks at Rutgers University Press, our agent, Deirdre Mullane, and the anonymous scientific reviewers who made so many excellent suggestions. We thank copy editor Bobbe Needham for improving the text, and Aisha Butler for creating the index.

We are extremely fortunate to have had so many capable and generous people give us their time, advice, feedback, and support. A group of them allowed us to use their wonderful images to illustrate our pages. Following is a partial list of those to whom we are indebted: Mark Adam, Frank Bond, Charlene Burge, Tom Cade, Peter Devers, R. David Digby, Nick Dunlop, Mick Ellison, Carl Engel, Tim Gallagher, Scott Haber, Isaiah James, Alan Johnson, Somchai Kanchanasut, Paul Keim, Mary LeCroy, Robert Leporati, Robert Mauro, Jonathan Meiburg, Karen Pidgeon, Anne Price, Bill Robinson, Elizabeth Schoultz, Jonathan Slaght, Alan Turner, and Ran Levy Yamamori. We intentionally leave out no one and we thank everyone who supported this project.

A tremendous amount of gratitude is due to the many ornithologists, falconers, and raptor enthusiasts who over the years and throughout the world have provided insight into the natural history of raptorial birds through their research, writing, and presentations, and whose efforts were vital to the continued existence of many species of birds of prey. We also thank the students and interested persons of all ages who continue to ask questions such as "How fast can a falcon dive?" It is imperative that we remain curious and informed about the natural world.

Coauthor Peter Capainolo would like to acknowledge all the family members and friends, too numerous to list individually, who gave him the encouragement and freedom to pursue his interest in birds over the years. Some of them deserve sainthood for putting up with live hawks in the garage, swans in the bathtub, and scads of dead, frozen specimens of all kinds in the kitchen freezer. He dedicates the book to his grandson Donovan Morant and his daughter Courtenay Lanner Thomas.

New York City
September 11, 2009

BIRDS OF PREY
HAWKS, EAGLES, FALCONS, AND KITES

Peregrine Falcon

Golden Eagle

Northern Goshawk

Broad-winged Hawk

White-tailed Kite

Swallow-tailed Kite

Merlin

Osprey

Short-tailed Hawk

Gyrfalcon

Ferruginous Hawk

Northern Harrier

BIRDS OF PREY
HAWKS, EAGLES, FALCONS, AND KITES

Red-tailed Hawk

Bald Eagle

Cooper's Hawk

Sharp-shinned Hawk

Prairie Falcon

American Kestrel

Red-shouldered Hawk

Harris's Hawk

Mississippi Kite

Swainson's Hawk

Crested Caracara

Rough-legged Hawk

Zone-tailed Hawk

BIRDS OF PREY
OWLS OF NORTH AMERICA

Barn Owl

Ferruginous Pygmy-Owl

Great Horned Owl

Flammulated Owl (gray phase)

Whiskered Screech-Owl

Barred Owl

Flammulated Owl (red phase)

Boreal Owl

Long-eared Owl

Elf Owl

Burrowing Owl

BIRDS OF PREY
OWLS OF NORTH AMERICA

Spotted Owl

Northern Saw-whet Owl

Eastern Screech-Owl (gray phase)

Northern Hawk Owl

Great Gray Owl

Short-eared Owl

Snowy Owl

Eastern Screech-Owl (red phase)

Western Screech-Owl

Northern Pygmy-Owl (gray phase)

Northern Pygmy-Owl (red phase)

How Fast Can a Falcon Dive?

ONE

Raptor Basics

Question 1: What is a raptor?

Answer: Raptors are birds that prey on animals that they either catch and kill themselves or find injured or already dead. The word "raptor" has its roots in the Latin *raptator,* meaning "to rob", and *raptore,* meaning "to seize and carry away." Raptors seize prey with their talons, either snatching fish from the water, striking birds out of the air, or pouncing on ground quarry such as mammals, reptiles, amphibians, and insects. Many birds feed on living organisms, for example, pelicans eat fish and robins eat worms, but when biologists use the terms "raptor" and "bird of prey," they are referring to the five to seven families of flesh-eating birds (depending on their point of view about classification) that include eagles, hawks, falcons, Ospreys *Pandion haliaetus,* vultures, and owls (see this chapter, question 4: How are raptors classified?). In addition to sharp claws or talons, a raptor has a sharp beak to tear flesh while feeding. These birds are equipped with keen eyesight and acute hearing, and most are strong fliers. They are usually brown or dull colored to provide camouflage. The word "raptor" is also applied to ancestors of modern birds— carnivorous, extinct dinosaurs such as *Velociraptor* and *Utahraptor* (see this chapter, question 3: What is the connection between "raptor" dinosaurs and modern birds of prey?).

Owls share these characteristics, but they are not closely related to eagles, kites, or hawks. Their relationship has been described as *convergent evolution,* in which animals not closely

Figure 1. This Cooper's Hawk *Accipiter cooperii* is a raptor on the hunt, its sharp eyes alert for prey. *(Photo courtesy of Richard Ettlinger)*

related to one another have evolved similar traits to adapt to similar environmental challenges. In this book, according to the existing scientific norm, we are including owls under the broad definition of birds of prey.

There are approximately five hundred species of birds of prey that are considered raptors. Over three hundred hunt during the day (diurnal), and the others hunt at night (nocturnal), and at dusk and dawn there are times when both types of birds may be active. Most owls are nocturnal, but, for example, Northern Hawk Owls *Surnia ulula* hunt during the day.

Question 2: Where in the world are raptors found?

Answer: With the exception of open oceans and the ice-covered continent of Antarctica, raptors are found on every continent as well as on isolated islands and ocean archipelagos. More hospitable environments generally support a broader range of prey species as well as a greater diversity of raptors, so tropical regions are teeming with life, including some of the most interesting raptors. Forest Falcons, genus *Micrastur*, found in the New World tropics, have facial disks that amplify their hearing, as does the Harpy Eagle *Harpia harpyja*, perhaps the most powerful eagle in the world. Small falcons such as Pygmy Falcons (genus *Polihierax*) hunt small insects in equatorial Africa, and Falconets (genus *Microhierax*) hunt insects in Asia. Some species have very limited distribution, like the Mauritius Kestrel *Falco punctatus* that is found only on the island of Mauritius in the Indian Ocean, and the Spanish Imperial Eagle *Aquila adalberti* that is found only on the Iberian Peninsula. Some species, like Peregrine Falcons *F. peregrinus*, are found almost everywhere.

There are raptors that monopolize the scarce resources found in very harsh habitats. Peregrine Falcons, Gyrfalcons *F. rusticolus*, Rough-legged Hawks *Buteo lagopus*, Golden Eagles *A. chrysaetos*, Short-eared Owls *Asio flammeus*, and Snowy Owls *Bubo (=Nyctea) scandiacus* inhabit the Arctic tundra, where there are few trees and the ground is frozen solid. In winter this biome (an ecosystem or geographical area containing specific plant and animal groups that are adapted to that environment) is in darkness twenty-four hours a day, which makes it difficult for any resident diurnal raptor to find prey. Golden Eagles and Rough-legged Hawks will feed on carrion, but all the tundra-inhabiting raptors take full advantage of abundant living avian and mammalian prey during the short summer breeding season. With the exception of the Gyrfalcon, most individuals of the remaining tundra species move south in winter to regions where prey remain abundant. Ptarmigan *Lagopus* spp. are grouse that remain in the tundra during the winter and provide a consistent food supply for the Gyrfalcon.

South of the tundra, the habitat changes to temperate coniferous forest or *taiga*. This biome consists of large tracts of evergreen trees of various species and ages and is often dotted with peat bogs (areas of decomposing dead plants, also called *muskeg*) and moorland (uncultivated hilly areas covered with coarse grass). A greater variety of raptor species occurs here than farther north, and many species do not migrate or are partial migrants, meaning not all individuals migrate. The taiga is the largest continuous forest in the world, and the preferred habitat of the Goshawk *Accipiter gentilis,* Sparrow Hawk *A. nisus,* and Sharp-shinned Hawk *A. striatus,* bird-hunting specialists. They share the forest with Common Buzzards *B. buteo* and Red-tailed Hawks *B. jamaicensis,* which feed mainly on small mammals. The taiga is also home to Gyrfalcons, Peregrine Falcons, and the robin-sized Merlin *F. columbarius.*

Deciduous woodlands, containing trees that flower and lose their leaves seasonally, are home to a wide variety of raptors; many breed and remain resident all year, although some, such as the Broad-winged Hawk *B. platypterus* and Eurasian Hobby *F. subbuteo* are migratory. Lanner Falcons *F. biarmicus,* Saker Falcons *F. cherrug,* and Peregrine Falcons are able to survive in some of the hottest, most arid regions on the planet. Open savannas, grassy areas with relatively high daytime temperatures, support many prey species for various harriers in the genus *Circus,* kites in the genus *Milvus,* and various falcons. The savannas also provide some vultures with the opportunity to feed on the carcasses of large grazing animals.

Some raptors nest, breed, and hunt in towns and cities. Peregrine Falcons, kestrels, Merlins, some vultures, and the Red-tailed Hawk (see color plate A), to name a few, have adapted quite well to urban landscapes around the world. Since 1992, a famous Red-tailed Hawk (nicknamed Pale Male) has been raising young in a nest on a twelfth-floor ledge above the entrance to a tall apartment building facing New York's Central Park. He has produced twenty-three offspring with a succession of mates, according to the many observers who follow his progress. His mate since 2002 has been nicknamed Lola. An Audubon census in 2007 reported thirty-two breeding pairs of Red-tailed Hawks nesting in New York City, and birdwatchers have reported hundreds more. A 2005 count listed eighteen breeding pairs of Peregrine Falcons nesting in the city. (See also chapter 8, question 5: What attracts raptors to live in cities?)

Question 3: What is the connection between "raptor" dinosaurs and modern birds of prey?

Answer: Fast-moving, predatory, bipedal (walking on two legs) dinosaurs, similar to the *Velociraptors* in the film *Jurassic Park*, were found in Asia, North America, and across the southern continents, dating from the Cretaceous period (about 144 to 65 million years ago). In 2005, a discovery in Argentina identified a 6-foot-long (2 meters) predator that lived about 80 million years ago. It had a hooked claw on each hind foot that appears to have been used to disembowel prey. This claw is a distinctive feature of dromaeosaurs (commonly called "raptors" or "running lizards"), the group that includes *Velociraptor*. Fossil evidence indicates that the size of early raptors ranged from that of a small dog up to about 30 feet long (9 meters).

With few bird fossils available to help explain the ancestry of birds, scientists have been forced until recently to speculate about their origins based on physical characteristics (morphology) and behavior. Discoveries of *Microraptor* fossils in China, reported in 2002, revealed that feathered species existed, a find that revolutionized paleontologists' ideas about how these creatures looked. Shannon Hackett of the Field Museum of Natural History in Chicago and colleagues published the results of an important genetic research project about birds in 2008. In that paper, ornithologist Rauri Bowie of the University of California, Berkeley, suggests that dinosaurs developed feathers from 65 to 100 million years ago and evolved into birds. Because dromaeosaurs share other characteristics with birds, it is believed that they were birds' earliest ancestors. In fact, some biologists conclude that birds are living dinosaurs. It appears that *Microraptors,* which weighed only a few pounds, were able to glide between trees on four-feathered "proto-wings." The discovery of additional species, unearthed in Mongolia in 1993 and described in 2007, contributed more evidence of the link between these small dinosaurs and modern birds. Theropod bipedal dinosaurs had similarities to birds in that they had similarly formed feet and

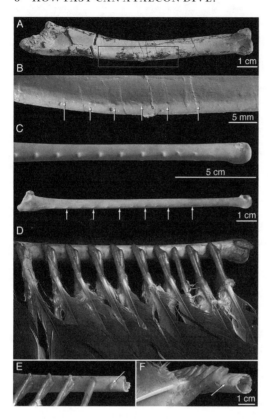

Figure 2. How feathers are attached: *A* and *B,* the carnivorous dinosaur *Velociraptor* apparently sported feathers on its fore-limbs, as evidenced by the feather quill attachment sites visible on the fossil ulna; *C* and *D,* attachment sites on the ulna of the extant Turkey Vulture *Cathartes aura; E* and *F,* how the feathers are aligned and attached to the ulna of a petrel, a group of modern flying birds. *(Photo courtesy of Mick Ellison)*

air-filled bones, laid and brooded eggs, and, in some instances, had feathers.

New research has made it possible to identify the color of feathers from prehistoric theropod dinosaurs. Analysis by Jakob Vinther and colleagues from Yale University of a feather found in Brazil shows that "most fossil feathers are preserved as *melanosomes,* and that the distribution of these structures in fossil feathers can preserve the colour pattern in the original feather . . . opening up the possibility of interpreting the colour of extinct birds and other dinosaurs." Melanosomes are organelles, self-contained parts of a cell (like organs in relation to the body) that contain the common brownish pigment melanin.

Dinosaurs and Birds of Prey

As part of my dissertation work on the group of meat-eating dinosaurs most closely related to birds, deinonychosaur theropods, I came across a very interesting feature on one of the specimens of *Velociraptor* I was studying: a series of regularly spaced bumps along the back edge of one of the lower bones of the arm, the ulna. It struck me as likely that these bumps represented quill knobs, the locations where secondary feathers attach themselves to the ulna (see figure 2).

Most people are probably familiar with *Velociraptor* as the villain of Steven Spielberg's movie *Jurassic Park*, but it is important to keep in mind that the animals in the movie were depicted as much bigger than then they actually were. In reality, they weighed only 7 pounds (around 15 kilograms). Finding quill knobs provided the first direct evidence for the presence of feathers in *Velociraptor*, and we can even count these bumps to estimate the number of feathers they represent—in this case, eighteen. Quill knobs are quite rare in nonavian dinosaurs and had not previously been reported in any other theropod.

Based on a wide range of fossil evidence, we know that the evolution of feathers preceded the evolution of birds. Indeed, the evolutionary sequence of feathers matches quite well with what we know about the developmental sequence of feathers. That is, the earliest feather forms that occur in the fossil record are also the earliest forms seen during the development of a feather. This is an outstanding case of biologists and paleontologists helping to illuminate each other's field. What is even more exciting is how recent fossil discoveries are changing the way we view all of dinosaur diversity, not just fossil birds and their closest dinosaur relatives. Recent discoveries in China from the Early Cretaceous period indicate that filamentous feathers, which were the evolutionary precursor to modern feathers, were likely present in all dinosaurs, not just the theropods.

(continued)

Dinosaurs and Birds of Prey, *continued*

The feathers that form the main part of the flight surface of the wing in living birds are not just embedded in the skin, as one might think. A follicular ligament runs from the quill of each feather to the bones of the hand and the ulna. The relationship between quill knob, ligament, and feather is quite complex. The contact of feather to quill and of quill to knob is not tip to tip, as one might expect, but instead is a pivot-point arrangement. The tip of the quill forms a cup that sits over the quill knob, and the follicular ligament is actually a sheet of tissue that surrounds and penetrates this knob. A series of parallel-running ligaments links the feather quills together. This provides an intricate support system for each feather and for the wing as a whole, and creates a cohesive surface that is acutely responsive to wing and wing-muscle movement.

The evidence of this ligament system in an animal like *Velociraptor* is important for understanding the evolution of feathers and flight. *Velociraptor* was too big to fly, so it remains unclear whether this arrangement of ligamentous attachments reflects retention from a flying ancestor, or if the system was later adapted by birds for a role in aerodynamic stabilization. To piece this puzzle together, scientists will continue to explore this soft-tissue morphology in modern birds and in future dinosaur discoveries.

—Alan Turner, research associate, American Museum of Natural History, and assistant professor, Department of Anatomical Sciences, Stony Brook University

Question 4: How are raptors classified?

Answer: Professional ornithologists have been engaged for centuries in an ongoing attempt to organize the approximately five hundred species of raptors into categories that reflect their relationships to each other and to other birds. Scientific names can be confusing, as birds were often named for people, although it was considered unprofessional to give a bird one's own name. Audubon named twenty-three species of birds for people he knew; Harlan's Hawk *B. j. harlani,* for example, was named for his friend Richard Harlan, a Philadelphia physician.

Scientists have traditionally used external features, skeletal characteristics, and behavior as taxonomic (organizational) criteria, and a number of species have been reclassified as new information is processed. Now that DNA analysis is possible and the genetic makeup of each species will eventually be completely mapped, the results will undoubtedly continue to shake up the accepted taxonomy until the evolutionary relationships among the species are clear. It still can be complicated to be sure about where a particular bird fits in—for example, there are at least fourteen subspecies of the Red-tailed Hawk in the United States, all having somewhat different coloration. Without genetic information, one cannot be positive about their relationships to one another.

The traditional classification of birds of prey is as follows:

Class—Aves (birds)
Subclass—Neornithes (modern birds)
Order—Falconiformes—(hawk-like birds)
    Family—Sagittariidae (Secretary Bird)
    Family—Pandionidae (Osprey)
    Family—Cathartidae (New World vultures)
    Family—Falconidae (falcons and caracaras)
    Family—Accipitridae (hawks, eagles, kites, harriers, and
        Old World vultures)
Order—Strigiformes—(owls)
    Family—Tytonidae (Barn Owls)
    Family—Strigidae (all other owls)

An example of traditional classification follows, using the Peregrine Falcon (see cover photo) as a subject. The Peregrine Falcon was first identified and described by John Tunstall in 1771. Its species name, *peregrinus,* which means "wanderer" or "pilgrim," reflects its proclivity to migrate long distances and its occurrence throughout most of the world.

Kingdom—Animalia (animals)
Phylum—Chordata (having a dorsal nerve cord at some
   point in the life cycle)
Class—Aves (birds)
Order—Falconiformes (hawk-like birds)
Family—Falconidae (falcons and caracaras)
Genus—Falco ("true" falcons)
Species—peregrinus

Hawks found throughout the world include short-winged hawks such as the goshawk, broad-winged hawks such as Swainson's Hawk *B. swainsoni,* and long-winged harriers such as the Northern Harrier *Circus cyaneus.* Other birds commonly referred to as hawks, like the American Sparrow Hawk *F. sparverius,* are actually members of the falcon family. Falcons are traditionally

Figure 3. Artist and naturalist John James Audubon described the Peregrine Falcon *Falco peregrinus* as the "Great-Footed Hawk" for good reason. The falcon's fully extended toes increase the surface area of the foot, making it easier to strike avian prey in midair. *(Photo courtesy of Nick Dunlop)*

Figure 4. A female Northern Harrier *Circus cyaneus* courses slowly over a salt marsh, its wings in a dihedral or *V* position. Adult males are smaller, with very light underparts and light gray upperparts. *(Photo courtesy of Richard Ettlinger)*

referred to as long-winged hawks by falconers. "Buzzard," a term that is sometimes incorrectly applied to vultures in the New World, actually refers to hawks in the genus *Buteo* or *Butastur*. Old World vultures, found in Europe, Africa, and Asia, are classified as belonging to the hawk family, Accipitridae.

Recent DNA research challenges the classification of some birds of prey, and it will remain in flux until more detailed genetic evidence replicates or contradicts the new data. For example, until the mid-1990s, all vultures were considered raptors order Falconiformes, based in part on their common behavior: they eat carrion, and their head and neck are devoid of feathers, perhaps to allow them to insert their head into a carcass with minimal risk of contamination. Old World vultures have strong beaks and powerful, hawklike, clutching talons, while New World vultures have thinner beaks and weak feet. Vultures also tend to urinate on their legs (urohydrosis), an unusual behavior they share with storks. Its purpose may be to cool their legs through evaporation (thermoregulation) or to destroy microbes that might adhere to them from decaying prey. In 1998, based on new DNA evidence, the American Ornithologists' Union officially grouped the New World vultures with storks and ibises, not raptors, and they were reclassified as Ciconiiformes, family Cathartidae. In 2007, Carole Griffiths of the American Museum of Natural History and colleagues challenged this classification and, based on DNA data, proposed a different reclassification. A large study reported in 2008 by Shannon Hackett of the Field Museum of Natural History in Chicago and colleagues also took a position on this controversy. Their analysis of the DNA found no affinity between Cathartidae (New World Vultures) and Ciconiidae (Storks) but "strongly supported placement of Cathartidae within the land birds (usually with Accipitridae)."

Studies of egg-white proteins in the 1970s by Charles Sibley and Jon Ahlquist, ornithologists at the Peabody Museum, Yale University, suggested a relationship between falcons and owls. Some scientists studying owls found indications that the phylogeny, or evolutionary history, of the owls needs official revision, but DNA work done at the American Museum of Natural History by Lisa Mertz, George Barrowclough, and Jeff Groth concluded that the traditional division of the owls into two families should remain intact. Interestingly, they found that, genetically speaking, Snowy Owls (see color plate G) that breed on the Arctic tundra are essentially big white Great Horned Owls *B. virginianus,* which are widely distributed in warmer climates. The Snowy Owls lack only the Great Horned Owls' feather tufts on

Taxonomy

Taxonomy is the science of naming living organisms, and the father of this field was a Swedish naturalist, Carl Linnaeus (1707–1778). Using Greek and Latin terms, Linnaeus popularized and refined a system of binomial nomenclature for describing and naming plant and animal species. A highly influential botanist, in his lifetime Linnaeus assigned unique names to thousands of species, so that no two organisms have the same combination of genus and species names. This eliminated the confusion caused by variations in common or colloquial names that frequently occurred before his system caught on. For example, the European Sparrowhawk is a very different type of raptor from the American Kestrel, but early colonists of the New World did not distinguish between the two and called the American Kestrel the Sparrow Hawk. The Linnaean system designates the European Sparrowhawk *Accipiter nisus* and the American Kestrel *Falco sparverius,* so that there is no difficulty understanding which bird is being described.

The first edition of Linnaeus's *Systema Naturae* was an eleven-page work, published in 1735 and paid for by a senator from Leiden in the Netherlands. By the time the tenth edition was published in 1758, it classified 4,400 species of animals and 7,700 species of plants. As more diverse and minute organisms were discovered, it became necessary to expand taxonomic designations to include bacteria, fungi, and other microbes.

The system of classification was originally based on observable shared physical characteristics. As new ways of detecting scientifically valid, observable characteristics have been developed, scientists began to classify organisms using DNA sequencing, taking into account their evolutionary relationships as well as their physical or morphological characteristics. This has resulted in the hierarchical relationships between some organisms being revised and understood in a more accurate way. Lindell Bromham of the Centre for Macroevolution and

(continued)

Taxonomy, *continued*

Macroecology at Australian National University, Canberra, wrote that Charles Darwin would probably have loved DNA, because "it solved one of the greatest problems for his theory of evolution by natural selection, . . . it gives us a tool that can be used to investigate many of the questions he found so fascinating, . . . and DNA data confirm Darwin's grand view of evolution."

their heads. The 2008 Hackett study gathered DNA from 169 species of birds and sequenced genes from fourteen chromosomes. Their data indicated that Falconidae and Accipitridae form distinct biological groups, or clades. that are descended from a common ancestor. Their findings suggest that falcons are more closely related to parrots and passerines (songbirds) than to the Accipitridae. Stay tuned!

Question 5: What are the differences between falcons and hawks?

Answer: Before analyzing the genetic material of an organism became possible, physical characteristics and behaviors were used to classify animals, and falcons and hawks have some apparent differences. Since ancient times, "true" falcons, genus *Falco*, were called "long-wings," and hawks, genus *Accipiter*, were known as "short-wings." The term "falconer" applied only to those who trained and flew long-wings. Although kestrels are actually falcons that hunt more like hawks, those who trained accipiters such as goshawks, sparrowhawks, or kestrels generally were of lower social status and were called "austringers," from the French *ostricier* and the Latin *accipiter*.

The common name "falcon" is used for several species of raptors, but only those in the genus *Falco* possess the full suite of characteristics of the birds we know as true falcons. With a few

Figure 5. Perhaps the most common large raptors in North America, Red-tailed Hawks *Buteo jamaicensis* can often be spotted soaring in lazy circles or perched on telephone poles or trees along roadways. *(Photo courtesy of Charlene Burge)*

exceptions, falcons have long pointed wings, their tails are fairly long and squared off at the tip, and most are powerful, swift flyers. The Peregrine Falcon is usually cited as being the fastest, especially in its stoop, or dive, when it zooms down to strike avian prey out of the air. Most falcons have long toes with a large surface area that helps them rake and kill prey. Kestrels can hover while searching for prey, a strategy different from the one most falcons employ. The shorter toes of kestrels are better suited for grabbing mice and insects in tall grass.

Falcon feathers are compact, their heads are small and blocky, and they have a less prominent supraorbital process (the bony protrusion above the eye) than many other raptor species have. The upper mandible of birds in the genus *Falco* has a toothlike projection, a tomial tooth, located just behind the pointed tip of the bill. The lower mandible has a notch that this structure fits into, and falcons use this equipment to bite the neck of their quarry for the coup de grace after subduing it with their talons. No other raptor species has this type of beak. True falcons do not build nests; they lay their eggs on cliff ledges, inside tree cavities, in the abandoned nests of other birds, on the ground, or on bridges or high-rise apartment building ledges.

Excluding owls, the term "hawk" is used to refer to all raptors other than falcons, but accipiters and buteos are the birds most people think of as hawks. These birds have broader wings and a more prominent supraorbital process. Accipiters have very long tails, and they fly rapidly, hunting birds in deep forests and woodlands. Buteos (soaring, broad-winged hawks and buzzards), harriers in the genus *Circus,* and eagles soar more often than falcons do and are more likely to prey on mammals. They also build stick nests on cliffs and in trees and are more anatomically diverse than birds in the genus *Falco.*

Question 6: Are all owls nocturnal?

Answer: Most people think of owls as creatures of the night, and this perception is mostly correct, as many owl species do indeed become active and hunt at night after their diurnal counterparts (hawks, eagles, and falcons) have finished hunting for the day. In temperate regions of North America, similar-sized Great Horned Owls and Red-tailed Hawks coexist but do not compete for food, because the hawks hunt during the day and the owls hunt at night. Barn Owls *Tyto alba* are probably the most nocturnal of all. They are highly adapted to finding and capturing quarry in almost total darkness and are quite elusive and secretive during daylight hours. Short-eared Owls hunt during the day, as do Snowy Owls (see color plate G) and Northern Hawk-Owls. Other owl species are crepuscular, meaning that they are active mostly at dawn and dusk.

There are four nocturnal Asian fish owls, genus *Ketupa,* and three African fishing owls, genus *Scotopelia.* They live in a wide variety of environments and have specially adapted feet that are completely different from those of other owls. Their feet are covered in scales and are similar to the feet of an Osprey, with a rough surface that enables them to grip slippery fish. Feathers on the Asian fish owls are not soft like those of other owls, and the wing feathers do not have the rough, sound-reducing margins like the wing feathers of other owls. They typically stand for hours perched beside a tropical river or stream, and when they spot a fish swimming close to the surface, they dive toward the water and grasp the fish with their feet, sharp talons extended. One of these owls, the Vermiculated Fishing Owl *S. bouvieri* found in Nigeria, feeds on clarias catfish *Clarias batrachus,* fish that have primitive lungs and must surface regularly for air. The threatened Blakiston's Fish Owl *K. blakistoni* is native to southeastern Russia and is probably the largest owl in the world. It hunts for fish by walking along the bank of a river, watching for a fish to come near the surface.

Figure 6. The endangered Blakiston's Fish Owl *Ketupa blakistoni* may be the largest owl in the world. Found only in northeast Asia, these owls live in pairs year-round and fish in unfrozen patches of icy rivers. *(Photo courtesy of Jonathan C. Slaght)*

Question 7: Are eagles the largest raptors?

Answer: The largest raptor and also the largest flying land bird in the Western Hemisphere is the Andean Condor *Vultur gryphus,* a New World Vulture (not an eagle), which can weigh over 30 pounds (15 kilograms) and has a wingspan of about 10 feet (up to 310 centimeters). The largest individual on record was 53 inches in length (135 centimeters). The California Condor *Gymnogyps californianus* is slightly longer from beak to tail, but it is overall not as large. The Cinereous Vulture *Aegypius monachus,* also known as the Eurasian Black Vulture, is the largest raptor in Eurasia, almost as large as an Andean Condor. Another very large raptor is the Eurasian Eagle Owl *B. bubo,* which has a wingspan of up to 6 feet (200 centimeters) and measures up to 30 inches in length (75 centimeters). Their weight ranges from 3.5 to as much as 9 pounds (2 to 4 kilograms).

While not the largest raptors, various eagle species have been used as cultural and national symbols because of their powerful and majestic demeanor. There is no exact scientific definition of an eagle, and the term is often used to describe very large raptors—one might describe eagles as very large hawks. Even Ospreys, family Pandionidae, are called Fish Eagles in some countries, although they are not closely related to eagles. According to a list on the Web site of the Eagle Conservation Alliance, there are about seventy-four eagle species, and many of their populations are in decline. The largest eagle on record is a twenty-nine-pound female Harpy Eagle, native to Central and South America, named after the half-woman/half-predator harpies in ancient Greek mythology (*harpazein,* meaning "to snatch" in ancient Greek).

Question 8: Are vultures the only raptors that eat dead animals?

Answer: Both Old and New World vultures are famous for feeding on carrion (dead animals) that either are relatively freshly killed or quite decomposed. The two groups are not closely related, and their similarity is thought to be due to convergent evolution. Vultures are opportunists, and although carrion is the preferred bill of fare of almost all vultures, when carrion is not available they will feed on small, living invertebrate and vertebrate animals that are not difficult to catch. A flock of Black Vultures *Coragyps atratus* in Mexico was observed following a stray dog, moving in to feed on its waste after it defecated. The Palm-nut Vulture *Gypohierax angolensis* is unusual in that it is a semi-vegetarian, preferring to feed on the fruit of the oil palm in Africa, although it will eat insects, fish, and even carrion if necessary to survive (see color plate B).

In the animal kingdom, there are many examples of species that are specialists, having a very specific diet and limited or no flexibility if their preferred food is unavailable. This is not the case with raptors, with the exceptions of the Snail Kite *Rostrhamus sociabilis* (see color plate F), which exclusively feeds on snails, and the Osprey, which eats mostly fish, and in some Eurasian regions also eats turtles. Many raptors that normally hunt live prey will feed on dead animals in cold weather when their usual prey is hibernating or not easily available. Golden Eagles have been seen feeding on dead deer in the winter, and other raptors feed on the remains of kills made by big predators like wolves or big cats.

Question 9: What is a buzzard?

Answer: Buzzards are medium-sized and large birds in genus *Buteo* and genus *Butastur*. In North America, *Buteos* are more commonly called hawks, but in other parts of the world they are described as buzzards. In the United States, "buzzard" can also refer to birds that are properly called vultures. The name was applied to vultures by early European colonists because the only soaring raptor in their countries of origin was a *Buteo*. The genus *Buteo* includes many familiar hawks, such as the Red-tailed Hawk, Swainson's Hawk, and the Common Buzzard.

Question 10: Which raptor is the smallest?

Answer: Native to the Sonoran Desert region of the southwestern United States, the 5-inch-long (11 centimeters), 1.5-ounce (about 50 grams) Elf Owl *Micrathene whitneyi* is probably the smallest raptor. It typically feeds on moths, crickets, scorpions, and other invertebrates. The forest-dwelling Northern Pygmy Owl *Glaucidium gnoma* at less than 2.5 ounces (73 grams) and just over 7 inches (over 18 centimeters) is a close runner-up. Its diet includes small rodents and reptiles as well as insects.

There are also some very small falcons. The African Pygmy Falcon *P. semitorquatus* at 3 ounces (100 grams) and 8 inches long (about 20 centimeters) is quite small. Its diet consists of insects, small reptiles, and some small mammals. The Black-thighed Falconet *M. fringillarius* is about 7 inches long (almost 18 centimeters). It is native to Thailand, Malaysia, Singapore, and Indonesia, and it feeds on insects and small birds. Color plate D shows a Collared Falconet *M. caerulescens,* another example of a small raptor.

Question 11: How long do raptors live in the wild?

Answer: Many or perhaps most raptors die in the first year of life before they even leave the nest, a time when they are inexperienced and vulnerable to predators. Small raptors like kestrels and Merlins are easy prey at any age, but if they survive their first two years they can live as long as eight years in the wild. During their first migration, some species have as high as 80 percent mortality because they are inexperienced, unskilled hunters. They may starve, and they run the related risk of being injured while approaching inappropriate prey. Estimates of the annual adult mortality for some raptor species ranges from 65 to 90 percent, according to Jean-Marc Thiollay of the Laboratoire d'Ecologie, Ecole Normale Supérieure, Paris.

If they survive their first year, medium-sized raptors like the Red-tailed Hawk can live ten to fifteen years. Large eagles, if they survive the approximately five-year transition to adulthood, can live twenty to thirty years in the wild, and they have been known to live much longer in captivity. The oldest known free-living eagle was thirty-eight years old, reported by ornithologist Helen Snyder in 2001. In 1982, Joel Welty reported records of a banded Osprey that lived for thirty-two years, a Golden Eagle that lived twenty-five years, and a Honey Buzzard *Pernis apivorus* that lived twenty-nine years.

TWO

Raptor Bodies

Question 1: How do male and female raptors differ?

Answer: Most male birds are larger and more brightly colored than females, but in some bird species, including birds of prey, there is an odd reversal of this rule, known as reversed sexual dimorphism (RSD). Females in some species, such as bird-hunting accipiters and falcons, are larger and more aggressive than males. In buzzards, eagles, Old World vultures, and owls, there is less sex-related size difference. Female Goshawks *Accipiter gentilis,* Cooper's Hawks *A. cooperii,* and European Sparrowhawks *A. nisus* are much larger than males and tend to be more aggressive, sometimes killing and eating their mates. Nestling female accipiters sometimes kill their smaller brothers while waiting for the adults to return with prey. Scientists are still not certain what role RSD plays in the ecology, biology, and evolution of raptors.

Most modern falconers use "falcon" to refer only to females and a completely different term, "tiercel," to refer to the male of any raptor species. Strictly speaking, these terms apply only to Peregrine Falcons *Falco peregrinus.* Adult males of some falcon species are more brightly colored than females, as is easily observed in male Merlins *F. columbarius,* which are very bluish on top, and in male American Kestrels *F. sparverius,* which have blue-grey wings. The females' wings are patterned and colored in muted tones like their upperparts.

Question 2: What do birds of prey eat?

Answer: Raptors are obligate carnivores, which means that they have evolved to eat meat and cannot survive on a plant-based diet. The one exception is the large Palm-nut Vulture *Gypohierax angolensis* native to sub-Saharan Africa and considered by various scientists to be an Old World vulture, a vulturine eagle, or a unique genus that seems to be intermediate between fish-eagles and vultures (see color plate B). Its preferred food is the nut of the oil palm, which it typically eats hanging upside-down in a tree, holding the fruit in its feet. Individuals have also been observed eating locusts, crustaceans, frogs, live or dead fish, and other carrion. They have occasionally been known to take domestic poultry, but if given a choice, they are said to prefer their vegetarian diet.

There are many kinds of small, medium-sized, and large raptors; some feed on small mammals, others feed almost exclusively on fish or other birds, and some seem willing to eat almost anything. Many smaller raptors such as kestrels, falconets, and some small species of owls and hawks eat small rodents and insects. The medium-sized Swainson's Hawk *Buteo swainsoni* is sometimes called a Grasshopper Hawk because it feeds on grasshoppers, crickets, locusts, and dragonflies—as many as a hundred insects in a day during the winter. During the metabolically costly breeding season, it eats heartier fare—mostly ground squirrels and other small mammals. The Eurasian or Steppe Buzzard *B. buteo vulpinus* forages on the ground for food, hunting for rabbits, field mice, small lizards, and snakes. It also attacks smaller birds in flight and eats flying insects like butterflies, bees, and locusts.

Jerry Olsen of the University of Canberra and colleagues collected prey remains and pellets to compare what Peregrine Falcons ate during four breeding seasons in two time periods. During the first period (1991–1992), their main prey were European Starlings *Sturnus vulgaris,* but eleven years later (2002–2003) when the starling population had declined, they ate a great variety of small native birds, thirty-seven species in all. Although they prey primarily on birds, if birds are not available they have been known to hunt on the ground for insects, lizards, or small mammals. In a study of the diet of the New Zealand Falcon *F. novaeseelandiae* during the 2003 and 2004 breeding season, raptor biologist Richard Seaton and colleagues found that they preyed on the more abundant species, intermediate in size, and frequenting open habitat.

There seems to be a raptor for every kind of prey, and some species are physically adapted to be very specialized in their diet. These specialists are not found where their prey is not abundant. The Snail Kite *Rostrhamus sociabilis* has a beak shaped in such a way that it can dine only on snails (see color plate F). Snake Eagles, genus *Circaetus,* live almost entirely on snakes. The Secretary Bird *Sagittarius serpentarius,* native to Africa, is a tall bird with long legs and short toes. It pounds snakes to death with its feet and swallows them whole, and it also takes other prey (see color plate C). Fish are the favorite prey of the so-called fish or sea eagles, such as the Bald Eagle *Haliaeetus leucocephalus*, African Fish Eagle *H. vocifer,* and White-tailed Sea Eagle *H. albicilla.* The Osprey, also called a Fish Hawk *Pandion haliaetus,* usually survives almost exclusively by capturing and eating fish. Barbed, spiky pads on the soles of its feet help the Osprey hold on firmly to slippery fish that are generally carried with the head forward in order to minimize wind resistance that would slow the bird's flight.

Many species migrate to find their preferred prey when cold weather in their home range causes seasonal scarcity. Small, insect-eating falcons like the Eurasian Hobby *F. subbuteo,* the Red-footed Falcon *F. vespertinus,* and the Lesser Kestrel *F. naumanni* make a long migration in winter from Europe to Africa to take advantage of the abundant insect population in the tropics. Sometimes there is an atypical movement of a large number of animals, called an irruption, that causes raptors to move from their home territory. If their prey disappears due to unusual climactic conditions or pesticide application, for example, an irruption of raptors may occur in an area where they are not normally seen.

Raptors that are generalists can survive on a wide variety of prey. The Red-tailed Hawk *B. jamaicensis* (see color plate A), the Golden Eagle *Aquila chrysaetos*, the Great Horned Owl *Bubo virginianus*, and the American Kestrel all eat a variety of animals, including wild birds, rodents, snakes, and even domestic fowl. Steppe Eagles *A. nipalensis* and Spotted Eagles *A. clanga* live primarily on small mammals and frogs, but in winter they prey on locust swarms. Raptors living in cities eat pigeons, rats, and mice. Black Vultures *Coragyps atratus* in the southeastern United States and Central and South America are found scavenging at dump sites, as are Black Kites *Milvus migrans* in parts of Europe and India. Coauthor Peter Capainolo once found an emaciated and weak adult Red-tailed Hawk with the remains of crickets and grasshoppers in its mouth. Upon examination, it was found that the bird had a healed fracture of a wing bone and was unable to fly. Although these hawks usually hunt small rodents, the bird had apparently been living on the ground for weeks and had resorted to feeding on insects in order to stay alive. (He recovered nicely after being given fluids and plenty of fresh meat. Because he would never be able to fly, he could not be returned to the wild, and he was placed in a large aviary at a nature center where, when we last checked, he remains alive and well.)

Old World vultures, family Accipitridae, and New World vultures, family Cathartidae, have extremely sharp beaks with which they tear into the hides, flesh, and internal organs of carrion (animals that have already been killed or have died of natural causes and have begun to decompose). They are less reliant on having the sharp talons and powerful grip needed to kill healthy prey, but they may prey upon wounded or sick animals that can be killed without using great force, and they may occasionally pick up and swallow small living invertebrates and mammals.

Question 3: How much does a raptor eat in a day?

Answer: Flying requires a lot of energy, and compared to most other animals, birds have much higher metabolic rates and typically are bigger eaters in proportion to their size. A small raptor may make a series of small kills during the day, while a larger bird may manage with one large kill every few days. During breeding season, raptors kill more prey to feed their young, and breeding usually coincides with the time of year when prey animals are abundant. The genetic mechanisms that determine how much energy an animal expends (and therefore how much it needs to eat) are not known, but recent research by Irene Tieleman of the University of Groningen, the Netherlands, and her colleagues suggest that mitochondrial DNA, inherited only from the mother, is an important regulator of energy metabolism in birds.

A bird of prey is usually not inclined to kill unless it is hungry, although satiated, "fed up" raptors may kill and abandon the prey, apparently just to stay in good form. On occasion they may kill and hide (cache) prey for later consumption. Falconers say that a hawk must be "in yarak," meaning in the mood to kill. This term is said to be of Arabic origin, possibly derived from the Persian *yaraki*, meaning "strength" or "ability," or from the Turkish *yarag*, meaning "readiness." Once a raptor has eaten its fill, it has little interest in hunting (see color plate A for an image of a raptor with a full crop). Prior to migration when they need to eat more to build up energy reserves so they can resist starvation, their flying-weight requirement limits the amount of fat they can store, and they instinctively balance storing fat against the need for a high level of agility and speed.

A Sharp-shinned Hawk *A. striatus* typically launches itself off a branch and pursues songbirds in flight, eating one quarter of its body weight each day. The amount eaten increases with the size of the bird, but, as is true in mammals, the ratio of body weight to food intake diminishes with size. So while a small, active accipiter like the Sharp-shinned Hawk consumes about 25 percent of its body weight each day, a medium-sized Peregrine Falcon consumes about 15 percent of its body weight in winter, less in summer. A larger Red-tailed Hawk consumes about 10 percent, a Golden Eagle consumes about 6 percent, and the largest raptor, an Andean Condor, weighing up to thirty pounds, eats about 6 or 7 pounds of meat a week, which amounts to only about 4 percent of its body weight each day.

The question of abandoned food or waste from a kill is relevant to quantifying how much raptors eat. They typically do not eat the entire prey animal, especially if it is large. They eat immediately on the spot and can carry off only the relatively small amount they can hold in their bill or talons while flying. Some raptors may return to a kill after a period of time and eat more of it, and some feed in pairs or in a group, in which case less is wasted. A portion of the weight of the kill is indigestible and cast up (regurgitated) as pellets. If preferred prey is not available, some raptors will kill an animal that is not their usual prey and then not eat it because they find it unpalatable. Leslie Brown, a Kenyan ornithologist, and Dean Amadon, at the American Museum of Natural History for most of his career, authors of the classic text *Eagles, Hawks, and Falcons of the World*, calculated that the waste factor for a Golden Eagle is about 20 percent, so it must kill five grouse to get the food value equivalent to four. As the ambient (external) temperature falls, heat loss increases and so does the energy birds require to stay warm, so they eat somewhat more. Birds in captivity that are inactive eat less, and older birds eat less than young birds do.

Secretary Birds (see color plate C) occasionally have been observed tearing a big kill into pieces and storing it under a bush to eat later, creating a cache of food. Unlike other raptors, owls do not have a crop—a storage pouch in the throat for extra food that can be consumed after the first portion of the meal has been digested. They are inclined to cache extra prey in a safe place so they can return to consume it later. Several smaller species of owls have high metabolic rates and need frequent meals, and they accumulate food as temperatures drop. The food stays fresh in the cold and may even freeze. As described by naturalist Cynthia Berger, when these owls are hungry, they simply sit on their frozen food until it thaws and becomes edible (see color plate G for three very different owl species).

Smaller falcons such as kestrels and Merlins are known to cache prey in rock or tree crevices, chipping nervously—"chip chip" is the sound they make, as distinguished from the "chirping" of songbirds—and transferring the prey item from foot to beak and back again while searching for an adequate place to cache it.

Question 4: How do raptors digest their food?

Answer: Modern birds have no teeth and limited jaw musculature, so they cannot chew their food. They swallow small prey whole and tear larger prey into manageable pieces that they then swallow. Some raptors pull out feathers or hair from larger prey so that they can digest the meat quickly and efficiently, extracting its nutritional value and remaining as light as possible in order to fly. Their weight is minimized by their hollow bones, by the absence of a urinary bladder and of teeth, and in some species by having one ovary that fails to fully develop.

According to research by Nigel W. H. Barton of Glasgow University, raptors' intestines differ in length by as much as 50 percent, related not to body size but to differences in foraging strategies. A raptor with a short, small digestive tract, like the European Sparrowhawk, Peregrine, or Hobby, digests its food rapidly. The Common Buzzard, European Kestrel, and Red Kite *Milvus milvus* have long, high-capacity digestive tracts and retain their food for a longer time. Studies by James K. Kirkwood of Langford House in the United Kingdom and by Andrew Tollan of the University of Canterbury in New Zealand explored the effect on metabolism of the physical and chemical nature of the raptor's diet, combined with the digestive efficiency of the bird. Energy assimilation efficiency increases when the ambient (external) temperature decreases, and after controlling the food intake, weighing the pellets and excretions, Tollan found that the Swamp Harriers *Circus approximans* in his experiments were consistent in the proportion of the different experimental diets they metabolized, and if they ate more they excreted more. Kirkwood found that captive kestrels and Barn Owls *Tyto alba* metabolized a similar portion of their food, suggesting that there is some consistency among species of raptors.

With the exception of owls, raptors have a crop, a loose sac in the throat that serves as a storage place for food that will be consumed later. Owls' food passes directly into their digestive system. Birds have a glandular stomach (a proventriculus) that produces enzymes, acids, and mucus. Digestion begins in that stomach, and then food moves into the muscular stomach, called the gizzard or ventriculus. In raptors, this stomach contains no digestive glands and serves only as a filter, holding insoluble parts of the prey such as bones, fur or feathers, and teeth. These indigestible parts, along with bits of grit or sand stuck to the prey and stones that some birds deliberately eat, all serve as rangle. To clean out the crop and ventriculus, rangle adheres to the undigested debris from the prey, and then the solid bits are cast back up out of the bird's mouth, coated with the fatty wastes and forming pellets (see this chapter, question 5: Can the species of a raptor be identified by its waste pellets?). Hawks generally eat fur, feathers, and bone (natural rangle), but falcons pick around that material and are usually more in need of roughage, so in captivity they are fed small rangle stones tucked into their meat. The twelfth-century emperor Frederick II of Hohenstaufen is reported to have fed his hawks wool as rangle.

The contractions of the ventriculus grind up the soft parts of the food, which then pass into the intestines, where secretions from the liver and pancreas provide digestive juices that process the food so it can be absorbed into the body. At the end of the digestive tract is the cloaca, a holding area for wastes from the digestive and urinary systems. The cloaca (Greek for "sewer") opens to the outside with a vent. Other than the Ostrich *Struthio camelus*, a big, strong, earthbound bird, birds have no bladder, since storing urine would weigh them down. Liquid waste is excreted as uric acid from the cloaca shortly after it arrives from the kidneys via a pair of tubes called ureters.

Question 5: Can the species of a raptor be identified by its waste pellets?

Answer: A few hours after having a meal, raptors that swallow their prey whole or in large pieces expel the indigestible parts of the prey that have been held in the gizzard or muscular

Figure 7. The composition of this pellet—feathers and a few bones—indicates it was produced by a hawk or falcon; the seeds are the victim's last meal. Mick Ellison found the pellet on his windowsill at the American Museum of Natural History. (*Photo courtesy of Mick Ellison*)

stomach (ventriculus) (see this chapter, question 4: How do raptors digest their food?). These remains—bones, teeth, fur, feathers, scales, and insect exoskeletons—are squeezed into a wet, slimy pellet that travels up the digestive tract and is eventually regurgitated when the digestive system has finished extracting nutrition from the remains. The bird stretches its neck, opens its mouth, and the pellet drops out without any retching or spitting. Depending on the species' typical diet and eating habits, it is possible to make an educated guess about which bird has cast up a dried pellet when one is found in the field.

The pellets of the African Crowned Hawk-Eagle *Stephanoaetus coronatus* were studied by Josh Trapani of the Smithsonian Institution and colleagues, who found that the contents of their pellets were distinctive enough that they might be useful in identifying fossil remains. These large raptors eat monkeys and even small antelopes, and the researchers found that the remains of their large-size prey, "crania, scapulae, and hind limb elements, are most likely to survive predation . . . and to be concentrated at Crowned Hawk-Eagles' nest sites."

Snake Eagles eat a snake whole and cast up very little in their pellets. The Secretary Bird is famous for its ability to kill snakes, even poisonous ones. Its long legs are covered with heavy scales that serve as an armor against snakebites, and the bird also uses its wings to shield its body and head from bites. Secretary Birds typically stomp or kick their prey to death, stooping to pick up the prey only after it has stopped moving and then swallowing it whole. They cast up large, sausage-shaped pellets that can be as long as 4 inches (10 centimeters).

Owl pellets contain more solid residue than those of other birds of prey, because their digestive juices are not strongly acidic and they tend to swallow their prey whole without plucking its feathers or fur. Owl urine typically stains the ground below their perches with "whitewash," which serves as a clue to the presence of their daytime roost. If you are trying to find an owl in a wooded area, look for whitewash and pellets near the base of a tree. Then look up—you might see an owl close to the trunk looking at you, trying to blend in with the foliage.

Question 6: Are raptors warm- or cold-blooded?

Answer: Raptors are warm-blooded animals (endotherms), meaning that they maintain a relatively level body temperature regardless of the ambient temperature, and they need to eat quite a lot to fuel their high metabolism. The need to maintain their flying weight limits the amount of fat they can store, and they actively regulate their energy reserves, balancing the need to keep their metabolism high against the risk of predation or injury due to excess weight from accumulated fat. Feathers help keep birds warm by trapping air next to the body, and owls have feathers even on their legs and toes that may protect them from injury as well as keep them warm. Boreal Owls *Aegolius funereus* that are native to colder habitats have feathers that are so efficient in protecting them against the cold that in warm weather they seek out cool roosts to avoid overheating. Burrowing Owls *Athene cunicularia* escape the summer heat by hiding underground (see color plate G).

Turkey Vultures *Cathartes aura* excrete liquid wastes onto their legs, possibly for its antimicrobial effect, and perhaps in hot weather to cool their legs as the liquid evaporates (thermoregulation). Like many birds, they also spread their wings to dissipate heat and to dry themselves. Zeev Arad, a biologist from Technion in Haifa, along with colleagues, studied the responses of captive Turkey Vultures to temperature variations. At low temperatures, the birds retracted the head and wings, and the skin on the head and feet was pale. At high temperatures, they extended the neck and wings, and the skin was deep red in color, indicating increased heat dissipation.

Jennifer Ward and colleagues at the University of Glasgow studied the response of Griffon Vultures *Gyps fulvus* to heat and cold. These vultures have large areas of bare skin, and they can change their posture to regulate heat loss by adjusting the extent to which their bare skin is covered by feathers. They can arrange themselves so only 7 percent of their bare skin is covered in hot weather, and up to 32 percent can be covered when ambient temperatures are low.

Panting to increase respiratory water loss is the primary method falcons use to dissipate heat, according to George A. Bartholomew of the University of California, Los Angeles and Tom Cade, formerly of Cornell University. Their work was followed up by James Mosher and Clayton White, formerly of Brigham Young University. They studied four species of captive falcons that normally live in different climates with a wide range in temperatures. They confirmed that falcons use the tarsus (leg) for thermoregulation, both by adjusting the amount of bare tarsus that is exposed and by controlling the circulation in that area to maximize or minimize heat loss. Owls also pant to cool themselves, and like vultures and other birds, they spread out their wings to allow cooling air to circulate underneath.

Question 7: What are the features of raptors' wings?

Answer: All raptors rely on their powers of flight to overtake prey, to maintain control of their territory, and to migrate long distances. But raptors' hunting and flying behaviors vary, and the morphological diversity in this group of birds is immense. Wing shape is linked with each species' ability to survive.

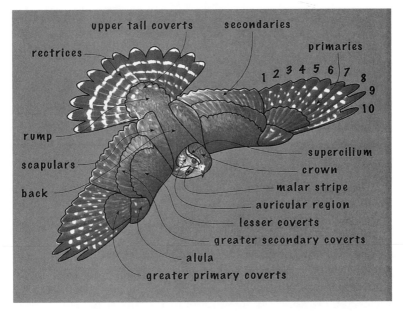

upper tail coverts   secondaries

rectrices

primaries

1 2 3 4 5 6 7 8 9 10

rump

scapulars

back

supercilium

crown

malar stripe

auricular region

lesser coverts

greater secondary coverts

alula

greater primary coverts

Figure 8. A raptor's dorsal topography and feather tracts are illustrated on a Merlin *Falco columbarius* in flight. *(Photo courtesy of Richard Ettlinger, modified by Jeff Groth)*

As in other birds, the skeletal anatomy of raptors' wings contains the basic elements of mammalian forelimbs. It consists of the upper arm bone (the humerus), the forearm (the radius and ulna), and all the finger bones fused together, with the exception of the thumb (see figure 8). The wing feathers grow in orderly tracts (rows), and the large secondary flight feathers of the wing are so deeply imbedded in the skin that they are associated with feather quill knobs, a row of evenly spaced knobs of connective tissue on the ulna where the tips of the secondary feather quills attach (see sidebar "Dinosaurs and Birds of Prey").

Long, notched primary feathers with asymmetrical webbing emanate from the fused finger bones. There are usually ten primaries in raptors, numbered P1 to P10, P10 being the outermost. Secondary feathers complete the complement of flight feathers and emerge along the forearm region of the wing. Large breast muscles power the downstroke of the wing, and smaller muscles and tendons lift the wing on the upstroke.

A raptor's hunting style varies in relation to its wing structure. Most true falcons, regardless of body size, have long, narrow, sickle- shaped wings with pointed tips. Most falcons fly with rapid, powerful wing beats, and some are among the fastest animals on earth. Kestrels have less dense pectoral muscles and fly weakly in comparison to other falcons, but they are able to hover in the air for long periods while searching for prey. Buteos, eagles, and vultures are masterful, soaring birds whose wings tend to be large and broad and more rounded in shape than the wings of falcons. This feature creates a large surface area under the wing that allows these birds to soar for hours using thermal currents of rising warm air. The combination of soaring and occasional wing flapping allows these birds to conserve a huge amount of energy. Short, broad, round-tipped wings make for extremely rapid bursts of flight in the forest-dwelling accipiters. These highly effective predators do not need to gradually build up speed; they can take off like a rocket from a sitting position.

The Red-tailed Hawk, Common Buzzard, and Great Horned Owl, although superb flyers, spend a great deal of time perched on limbs, ready to pounce on prey directly beneath them. The Secretary Bird strides along on the ground on its long legs, searching for small mammals, lizards, and snakes. Occasionally one will take on a venomous snake and pound it to death with its toes, all the while keeping its wings spread, apparently to confuse the snake and direct its dangerous strikes toward the primaries where they can do no harm. Other raptors, like the Northern Harrier *C. cyaneus*, fly low and slow, close to the ground in salt marshes and among the dune grass along barrier beaches. They hold their long wings in a *V* or dihedral position, as do Turkey Vultures. In comparison to rapid, prolonged, flapping flight, they conserve energy by pouncing from a perch, walking and flashing wings, and coursing slowly over terrain.

Question 8: Do raptors molt?

Answer: Molting is the periodic loss and regrowth of feathers, a natural process in all birds. Feathers are amazingly strong and resilient structures, but over time they become worn, frayed, broken, and faded, making them less useful, and so they are lost in a molt and replaced. The timing and sequence of molt is tremendously variable.

In general, raptors molt once each year, but not always completely. A newly fledged raptor has a fresh coat of perfect feathers, and in many species the color and pattern of the first year or "passage" immature bird differs from the coloration of an adult. The following spring, the young bird undergoes its first molt and replaces many, if not all, of the major flight, tail, and body feathers, and it begins to look more like the adult of the species. It may take several years for a raptor to molt completely into adult plumage; in some of the large eagles, condors, and large Old World vultures this may take as long as four or five years. Peregrine Falcons that breed in the Arctic and spend the winter in the southern hemisphere may start the wing molt in May or June, completing it in February or March when they are in the South.

Some birds, such as geese and ducks, molt all their flight feathers at once and cannot fly until they regrow. For a raptor such a total loss would mean certain death, since they depend on flight to capture prey. Thus raptors molt slowly and, in most species, symmetrically, so as not to impair their flight capability. For example, a falcon may molt P4 of the right and left wings at the same time and then not molt the adjoining primary until fully half of the new P4 has emerged. Tail feathers are molted in a similar fashion, which keeps raptors balanced for flight. With less need for balanced flight, vultures that generally eat carrion molt asymmetrically and in no particular sequence. While the molt sequences of many raptor species are known, others still need to be studied and documented.

Interestingly, raptors often feed on prey that are molting, shedding, newly born, or just fledged, suggesting that their weakness or inexperience makes them easier to capture, especially for a young raptor.

The Bald Eagle does not attain the conspicuous white head and tail plumage that gives it its name until it is at least four or five years old. Young eagles approaching maturity often seem larger than adults, and in fact they are. Their longer wing and tail feathers are more conducive to soaring on thermal air currents and to making longer, wandering flights, while mature eagles hold their territory, using and rebuilding the same nest (or a nearby alternate) for many years, and straying only to forage. With each successive molt, the Bald Eagle's wing feathers become somewhat shorter and narrower and the tail feathers become shorter, resulting in aerodynamic properties more suited to their settled life at maturity.

Question 9: What is "preening"?

Answer: Birds keep their feathers in good condition by preening, which involves stroking the feathers with the bill or talons, smoothing the surfaces, repairing small splits in the feathers, rearranging them into their normal configuration, and cleaning them of debris and parasites. A bird may preen with a nibbling action, moving its beak along the length of a feather from base to tip, or with a firmer action, gripping the feather and pulling it through the bill. Mated pairs of raptors preen each other, and parents often preen their nestlings and fledglings.

While preening, the bird presses its beak periodically against the uropygial (oil) gland above the base of the tail, releasing oil to spread over the feathers so that water will bead and run off. The oil also contains antibacterial and antifungal agents. Because ultraviolet light converts substances in the oil into vitamin D, when the bird preens after the oil has been exposed to sunlight, it ingests the vitamin D, which helps it to process calcium.

Birds pay special attention to the flight feathers of the wings and tail, but raptors also preen the smaller body or contour feathers. Since they cannot preen their head feathers (imagine trying to lick the top of your head), raptors rub the sides of their head against their shoulder. Barn Owls have a special comblike structure on the middle talon that they use just for preening and grooming their head. With few, if any, head feathers, vultures rely on a different kind of hygiene (see chapter 5, question 3: How do raptors defend themselves?). Raptors also raise, rapidly shake, and lower their feathers as part of feather maintenance, a procedure known to falconers for centuries as "rousing." Rousing and preening are considered signs that a bird is in good physical and emotional health.

Question 10: What is special about raptors' feet?

Answer: Like most birds, birds of prey have four toes on each foot, usually three facing forward and one facing to the rear. They have sharp talons and a strong grip, and they typically pounce on their prey, piercing the animal's body with their talons and, if necessary, delivering a killing bite. Most falcons have exceptionally long toes, and their rear talon is particularly powerful. The long toes increase the surface area of their feet, making it easier for them to strike birds out of the air. Falconers wear a thick leather glove, usually on the left hand, to protect their hand from the sharp talons and strong grip of the raptors they handle.

Owls have two toes facing forward and two facing to the rear. The outer toe can swivel forward or backward as needed, allowing for a versatile grip that is very powerful. Their toes automatically spread when they straighten their legs to strike a target, and when their feet hit the prey, the legs bend and the toes grip automatically, driving the talons forcefully into the prey. Owls also use this auto-lock feature to carry prey or to grip a perch without having to use energy to contract their muscles.

The Osprey, a specialist in catching fish, has feet specially equipped for this slippery task. The undersides of its feet are rough and scaly, which helps it hold firmly to the fish. The Osprey's grip is so vicelike that it sometimes cannot let go of a large fish, and there is a popular anecdote about a very large carp caught in a lake in Saxony with the skeleton of an Osprey still attached by its claws. (We have not been able to obtain a copy of the photograph of the carp allegedly published in the March 20, 1958, issue of the British field sports magazine, the *Field*.)

Question 11: How well do birds of prey hear?

Answer: We know that raptors have adequate hearing because most are quite vocal, using calls to communicate with their partners. They can respond to a trainer's whistled signal and are likely to fly away at a loud sound nearby. The relationship between hearing ability and hearing-related behaviors has been studied in relation to inner-ear bony anatomy by Stig Walsh of the Natural History Museum, London, and colleagues. They found a positive relationship between the length of the endosseous cochlear duct and the hearing ability of fifty-nine bird and reptile species; better hearing correlated with complex vocalizations, pair bonding, and living in large groups. Birds are able to hear sounds between about 1 kHz to as high as 10 kHz, but their hearing tends to be less sensitive at both the high and low ends of the range. Humans, in comparison, can hear sounds ranging from about 0.02 kHz to about 17 kHz.

With a few exceptions, birds of prey hunt by sight, although some species that hunt small ground animals can detect very faint prey sounds and can discriminate between very close frequencies. Hawks can recognize one another by their individual voices, and they use loud and varied cries as part of their courtship displays. Established pairs of hawks are commonly observed using a variety of sounds in their interactions with one another.

Barn Owls and some other owls are unique among raptors in that they have an asymmetrical bony ear structure or asymmetrical soft parts associated with the ear. This provides them with very acute hearing and allows them to pinpoint the location of mice and other small prey animals in total darkness. The soft feathers on their wings enable them to fly silently so they can stealthily approach prey without the sound of their wing beats either alarming the prey or interfering with their own ability to hear. Birds of prey that fly low to the ground and are partially nocturnal tend to have more acute hearing to supplement their vision. Forest falcons, genus *Micrastur,* and harriers, genus *Circus*—genera that hunt in areas where foliage impedes their vision—have a facial disc, a sound-reflecting ruff of skin and feathers, and larger ear openings than is usual among raptors. They attack prey that they can hear but not see in the dense vegetation. Great Grey Owls *Strix nebulosa* plunge feet first into snow to catch rodents that they cannot see but detect with their acute hearing.

Northern Harriers were studied by William Rice at Oregon State University, where he tested their hearing and compared it to that of Red-tailed Hawks, American Kestrels, and owls. Northern Harriers were able to detect synthetically produced squeaks of voles without visual or olfactory cues from a maximum of 3 to 4 meters, while a Barn Owl could detect the squeaks from 7 meters. Northern Harriers share with owls the facial disc and the habit of foraging close to the ground. This is in contrast to most falcons, hawks, and eagles, which typically hunt from much greater heights and rely almost exclusively on vision.

Question 12: How well can raptors see?

Answer: Birds of prey have the most highly evolved eyesight of all living animals. If you had eyes like a kestrel, you could probably read a newspaper from 25 yards away. With eyes like a hawk, if you were a Red-tailed Hawk, you could see a mouse from as far as a thousand feet in the air. And if you are eagle-eyed like a Golden Eagle, you could spot a rabbit from a mile or two away. A Barn Owl's night vision is the equivalent of your being able to see a mouse from a mile away by the light of a match, and this owl's extraordinary night vision is supplemented by unusually acute hearing (see this chapter, question 11: How well do birds of prey hear?).

Vision is a diurnal bird's most important sense, and these birds depend on precise and subtle visual discrimination for their survival in the wild. Raptors' eyes are on the front of the

Figure 9. Eyes at the front of its head, typical of raptors, and a laterally compressed bill give the Steller's Sea Eagle *Haliaeetus pelagicus* the advantage of full frontal vision. *(Photo courtesy of Jeope Wolfe)*

head, like the eyes of a human, with overlapping binocular fields of vision and good depth perception. This placement contrasts with that of birds and other animals that are prey—their eyes are on the sides of the head, which provides wider coverage for spotting predators but gives them monocular vision with poor depth perception and limited ability to judge distances.

Experiments over the years to investigate the visual acuity of different species of raptors have used various instruments and measurements—translucent plastic goggles with interchangeable lenses placed on the bird, pattern electroretinograms (using electrodes in contact with the cornea to record retinal responses), and frozen ocular sections of sacrificed birds. In most experiments, the bird is trained to respond to one stimulus in the presence of others that are different, and a correct choice earns a food reward. Once the bird has learned the desired response, the experimenter manipulates some property of the stimulus, such as how far away it is, until the bird can no longer discriminate. (See "The ultimate guide" in References for a fascinating video of an experiment by an Australian scientist using two video screens and a food reward to study the vision of a Golden Eagle.)

All vertebrates have the same basic eye structure (pupil, iris, retina). Birds' eyes are very large, larger in proportion to their body size than the eyes of other vertebrates—the two eyes together can weigh as much or more than the bird's small brain. An eagle's eye, for example, is physically larger than a human eye, and even a Snowy Owl *Bubo (=Nyctea) scandiacus*, typically weighing less than five pounds, has eyes the size of those of a 200-pound man. Because the eyes take up so much space in the bird's head, moving the eyes around in their sockets is not a real option because there is not enough room for the muscles required. This results in typical birdlike head movements to get an object in focus. To swivel its head rapidly and see all around by turning only its head, an owl has fourteen vertebrae in the neck, seven more than most birds. In contrast to owls' 360-degree visual field, human's visual field is slightly over 180 degrees (to check, look behind you by turning just your head).

The retina is a screen at the back of the eye onto which an image of what you are looking at is projected. A large eye means a large retina, and the retina of a bird can also be up to twice as thick as that of mammals. A huge number of light-sensitive cells—rods and cones—cover a bird's retina, and more sensory cells mean better visual acuity. Rod cells on the retina work well in low light and are sensitive to black and grey tones, and they register the shape of an object. The human eye contains about two hundred thousand rod cells per square millimeter of retina, while an owl's retina has about a million rods in the same space. Cone cells require high levels of light to operate and are sensitive to bright light and colors. Cone cells are often topped with an oil droplet that heightens the contrast of colored objects and also acts as a haze filter.

In addition to the basic structure they share with all vertebrates, the eyes of falcons, hawks, and eagles have two tiny pit-like regions, or foveae, on each retina (*fovea* is Latin for "small pit"), and it is in these areas that cone cells are the most densely packed. Many animal species lack true foveae (they are afoveate), and primates are the only foveate mammals. The pit- or funnel-shaped fovea refracts light, enlarging a viewed image and enhancing its focus. Humans have only one fovea; when we focus directly on an object the image falls on the fovea and we see it clearly, but our peripheral vision is weak. Bifoveate raptors (those having two foveae) use the second fovea to help track moving prey, making it easier when hunting in open country to judge the speed and distance of potential prey.

Most diurnal raptors have one deep and one shallow fovea. The deeper pit, also called a nasal or central fovea, provides a forward line of sight that extends about 45 degrees to the right or left of center and is adapted to perceive fine details in distant objects. To get a distant image to fall on this fovea in the retina so it can be seen with maximum clarity, the bird turns its head sideways to see an object straight ahead. The second, shallower fovea, sometimes called the lateral or temporal fovea, covers a range of about 15 degrees to either side of the center and has lower visual acuity but is important for stereoscopic or three-dimensional binocular vision. Vance Tucker of Duke University, who has studied the behavior of several species of raptors, observed that they all typically move their head into a similar position when looking at an object. When the object was nearby, they looked straight ahead at it, apparently relying on the shallow fovea for three-dimensional vision. As the distance to the object increased, they looked sideways at the object 80 percent or more of the time, apparently using the more acute sideways vision in the deep fovea to obtain maximum sharpness. Katherine Fite and Sheila Rosenfield-Wessels at the University of Massachusetts found considerable variation in the location and configuration of the foveae in nine avian species with a wide range of eye sizes and ecological habits. The relative strength of the foveal areas depends on their visual purpose so, for example, the various species of bifoveate birds known as kingfishers, order Coraciiformes, use the temporal fovea while diving underwater for small fish and the nasal fovea while in the air.

Carrion-eating Andean Condors *Vultur gryphus* (see color plate B) and Black Vultures *Coragyps atratus* have only a nasal fovea, consistent with their having less need for acute vision because of the immobile nature of their typical prey. Owls have only temporal foveae, probably related to their more frontally placed eyes and greater binocular field. The owl retina is also unique in that its fovea contains primarily rods instead of cones, an adaptation to owls' nocturnal lifestyle. Owls' eyeballs are longer from back to front than from side to side, so they function somewhat like binoculars to provide enhanced distance vision—their vision is excellent even without that second fovea.

The upper and lower eyelids of birds are controlled by smooth, involuntary muscle, and birds usually close the lids only when asleep. Most birds and some other vertebrates have a third transparent or translucent eyelid, a nictitating membrane, which sweeps across the eye's surface, keeping it moist and clean under most circumstances. Striated, voluntary muscles control the nictitating membrane, which is used throughout the day not only to keep the eye clean but also to protect it from branches and thorns and from flailing prey. Hawks have a projecting bony shield above the eye, a supraorbital process, that provides further protection for the eye. For raptors like vultures that feed on carrion, this extra protection is less essential, and some species may lack the bony eye shield.

In 1996, Finnish scientists led by Erkki Korpimaki at the University of Turku published the results of research demonstrating that at least some birds of prey can see a color spectrum that includes ultraviolet light. Eurasian Kestrels *F. tinnunculus* prey on small rodents that are often found in large grassy habitats, and the rodents' urine leaves stains that appear black in ultraviolet light. In the laboratory, the researchers confirmed that the kestrels could see the ultraviolet urine stains left by mice and voles and that they could discriminate between active and abandoned urine trails. This gives them an evolutionary advantage because they can visually screen large areas of vegetation and focus their hunting on areas where prey was recently active.

The human eye adjusts slowly to changes in light conditions, and the muscles that control how much light is admitted are smooth, involuntary muscles. It takes more than thirty minutes in the dark for our eyes to completely adapt and for our nocturnal vision to function at its highest level. Even a momentary burst of light causes regression to night blindness and once again a relatively long period of time is required for the retina to readjust. In contrast, birds control these adjustments with striated, voluntary muscles, so they can adapt quickly to rapidly changing light conditions. The hawk is particularly adept at making such adjustments, but raptors that do not pursue active prey, such as vultures, are less so. Fewer than 3 percent of birds are principally nocturnal, and naturally their vision is best adapted to dim light. Because most owls hunt at night, their pupils can open very wide to admit the maximum amount of light, and each pupil adjusts independently to the amount of light on its side of the head.

Many raptors have dark head patterns that probably reduce glare, like the dark grease under the eyes of a ballplayer. There are other special adaptations that make birds of prey such successful hunters, but even this cursory exploration shows that their visual systems play a large role.

Question 13: Do raptors have a keen sense of smell?

Answer: In most birds, the part of the brain related to smell (the olfactory lobes) is quite small, and birds rarely can detect airborne scents. However, some birds have scent-detecting cells in a cavity in the nostril that opens directly into the mouth, and so they may be able to smell something that they are holding in their bill. Nick Fox, raptor biologist, breeder, and falconer, has observed that various hawks pick up and reject meat that they have held only in the tip of the bill in such a way that it seems unlikely they have tasted it, so at close quarters they may have some sense of smell through the mouth or nares.

Turkey Vultures may be the exception. They are said to have a highly developed sense of smell. Two studies in which researchers hid dead animals and Turkey Vultures flew directly to the cache provide evidence that their sense of smell is important for finding food. Although they apparently prefer fresher meat to meat that is rotten, in experiments conducted by David Houston, an ornithologist at Glasgow University, the birds could not find dead chickens he had hidden under leaves until the second day, when they were giving off a stronger odor. Houston was involved in another study in Colombia with Luis Gomez of Santa Fe de Bogotá and colleagues, where they found that Greater Yellow-headed Vultures *C. melambrotus*, using their sense of smell, were able to locate 63 percent of hidden carcasses, while mammalian scavengers found only 5 percent.

William McShea of the Smithsonian Conservation and Research Center had similar results using road-killed deer covered with hay, which Turkey Vultures found within a day or two. Biologist Bill Robinson, in a personal communication, maintains that he has never seen any convincing evidence that his captive Turkey Vultures are able to find food using their sense of smell, and he and some colleagues are skeptical about the Turkey Vulture's touted olfactory ability.

Vultures' circling over a carcass may mean they are trying to determine if it is still fresh enough to eat. Black Vultures are known to fly very high, observing low-flying Turkey Vultures that are searching for carcasses. When the Turkey Vultures descend on a find, the more aggressive Black Vultures try to monopolize the meal. In the American tropics, the spectacular-looking King Vulture *Sarcoramphus papa* (see color plate B) follows the Turkey Vultures as well, and it is big enough to chase the Turkey Vultures *and* the Black Vultures away from a carcass.

Kenneth E. Stager reported that in 1938 the Union Oil Company discovered that Turkey Vultures could detect natural gas leaks in its pipelines when ethanethiol was injected into the lines. The birds would circle over a leak, apparently attracted by the odor of the additive, which resembles the odor of carrion.

THREE

Raptor Behavior

Question 1: How intelligent are birds of prey?

Answer: An animal can have good vision and a good memory and not be particularly intelligent. We know that raptors generally have good memories, as is apparent when trained birds recognize former owners after a long absence. Raptors have relatively large eyes in proportion to their body, and their vision is superior to that of most other birds and mammals. Large eye size means the eye can admit more light, and the retina will be larger with room for more rods and cones to provide greater visual power. A correspondingly larger area of the brain is dedicated to efficiently process the flood of visual information.

We know that the relative size of an organ has some relationship to its functional significance, and a larger brain is associated with increased cognitive skills and behavioral flexibility. Lazlo Garamszegi of the Université Pierre et Marie Curie in France, along with colleagues Anders Pape Møller and Johannes Erritzoe, examined the relationship between eye size and brain size in birds, and the correlation of eye and brain size with the demanding visual tasks of capturing actively moving prey and nocturnal activity. They conducted postmortem examinations of 2,716 birds and analyzed their measurements in conjunction with data obtained from handbooks and field guides on how they capture prey. Their results statistically demonstrated a positive relationship between eye size and brain size, and a positive relationship between eye size and complexity of prey-capture technique.

The Harris's Hawk *Parabuteo unicinctus*, a North American raptor, uses a coordinated and cooperative hunting strategy when going after prey such as fast-moving small birds and large jackrabbits. This is probably the most advanced form of cooperative hunting known among birds, and it improves their capture rate and allows them to dispatch prey larger than themselves. Groups of from two to seven individuals, usually a pair and their offspring, use several different methods to hunt. They generally are found in desert and savannah habitats with dispersed cacti or small trees on which they can perch to spot prey. In one technique, several hawks pounce on a prey item in an open area. Another method is for one hawk to flush prey from cover while other hawks make the capture. Hawks take turns in the lead during extended chases after prey, so that if one hawk misses or tires, another can take its place and the chase will not flag. Alternatively, a second hawk will come from the opposite direction to intercept the prey. These hawks have become popular for falconry because of their cooperative nature and the ease with which they bond with a trainer and work in tandem with dogs and other trained Harris's Hawks.

Question 2: How fast can a raptor fly?

Answer: Raptors are famous for their powers of flight, and most raptors are strong fliers, although there are some exceptions like the Secretary Bird *Sagittarius serpentarius*, which hunts while walking. Speed is essential to surprise and overcome prey that is fleeing at top speed. Red-tailed Hawks *Buteo jamaicensis* usually soar in lazy, slow circles while searching for prey, but they can drop quickly from 1,000 feet (over 300 meters) to snatch a rodent or rabbit. The small Sharp-shinned Hawk *Accipiter striatus* and European Sparrowhawk *A. nisus* zip through dense forest cover at speeds of 40 miles per hour (64 kilometers per hour) or faster in pursuit of prey. The large Golden Eagle

Bird Strikes

"It was the worst sickening, pit-of-your-stomach, falling-through-the-floor feeling I've ever felt in my life," said the Airbus 320 pilot shortly after it crash-landed. "I knew immediately it was very bad." In January 2009, he had just taken off from New York's La Guardia Airport when he noticed a line of birds on the right side of the aircraft. An instant later the windscreen was filled with birds, and his first instinct was to duck. Then there was a thud and both engines stopped, and he smelled burning birds. He landed the commercial jetliner smoothly in the Hudson River adjacent to midtown Manhattan, and all 155 people on board were rescued in short order, thanks to the boats and helicopters that converged on the plane from both nearby shores.

A bird strike was confirmed as the cause of the engine failure, and feathers and bird remains recovered from the engines were sent to the Smithsonian Institution so the species could be identified. The speculation was that a local flock of Canada Geese *Branta canadensis* was involved. When the press interviewed coauthor Peter Capainolo shortly after the accident, he explained: "If they were geese, the birds would have been large enough to do considerable damage." Canada Geese weigh about nine pounds (over four kilograms). Subsequent isotopic analysis of feathers recovered from the accident confirmed that the culprits were migratory geese on their way south from Canada (the chemical composition of feathers reflects the type of grasses that a bird was eating when the feathers grew, and in this case it was Canadian grass).

Planes flying out of the New York area are at high risk for bird strikes because all three major airports (and several small ones) are close to active coastal wetlands, waterways, wildlife preserves, and migratory flyways. In the last nine years in the United States, almost five hundred fixed-wing airplanes have collided with birds, according to the Federal Aviation Administration, often resulting in damage, emergency landings, and aborted takeoffs. A report of a joint United States and Canada

Bird Strikes

Bird Strike Committee found that between 1990 and 2005 in the United States, 370 reported bird strikes to civil aircraft involved helicopters, and half the helicopter reports indicated damage. A similar problem exists worldwide.

There have been attempts to reduce the bird population adjacent to airports by removing shrubbery that attracts prey and provides cover for the birds. Other efforts have been made to scare them away, euthanize them, and even shoot them. A $3 million, five-year program run by Falcon Environmental Services at Kennedy Airport uses teams of falconers to control the birds. According to Mark Adam, the company's president, this is an extremely effective and environmentally conscious use of ancient techniques to address a modern problem.

Most nuisance birds are active during the day, and each airport falconer patrols during the daylight hours with up to five falcons that he or she is assigned. When birds are spotted, falcons are taken to the area and one is released. Just the silhouette of these fierce hunters in the sky causes the nuisance birds to flee. When the birds have gone, the falconer swings a lure and the falcon returns to the glove.

So the next time you see a flock of roosting birds suddenly take off as if controlled by one mind, look to the skies. Chances are there's a falcon or another raptor up there, and the birds are instinctively fleeing for their lives.

Aquila chrysaetos at first glance does not appear capable of rapid flight, but in hot pursuit of a jackrabbit or a goose, it can attain a speed of 60 miles per hour (over 96 kilometers per hour). Golden Eagles are also amazingly adept at making rapid turns and twists in mid air.

In a sustained vertical dive, the Peregrine Falcon *Falco peregrinus* is probably the fastest bird on earth (see this chapter, question 3: How fast can a falcon dive?), but their cruising speed during migration has been measured at between 40 and 60 miles per hour. Airplane pilots have reported observing raptors at cruising altitudes over 3,500 feet (over 1,000 meters), and airplanes taking off or landing periodically strike them (see sidebar "Bird Strikes").

Question 3: How fast can a falcon dive?

Answer: A Peregrine Falcon usually captures flying birds by rising high, sometimes thousands of feet above the prey, and making a spectacular dive or "stoop" aimed at the quarry. Fighter pilots flying close to diving peregrines have estimated their speed at up to 200 miles per hour (322 kilometers per hour), and recently Ken Franklin, a skydiving Canadian falconer, jumped out of a plane with his trained peregrine and clocked it diving at 242 miles per hour (389 kilometers per hour). Peregrines probably do not routinely dive after prey at such high speeds, but their ability to do so is an amazing testament to the powers these magnificent raptors possess.

Extreme performances like this occur in situations that have important fitness consequences—besides an attack on elusive prey, they can occur in escape behavior and in courtship displays. For example, the tiny Anna's Hummingbird *Calypte anna* flies at almost double the maximum speed of the Peregrine Falcon for periods of 0.3 seconds during its courtship display dive.

Slow-motion videos show that the falcon spreads its toes open to grab the prey at the moment of contact, but because of the high speed at which this occurs, it instead rakes the prey, usually with the hind talon. Because this happens so fast, it can look like a closed-foot strike.

The rush of wind when a falcon is in a high-speed stoop could dry up the normally dilute tear film on the surface of its eyes, compromising its vision. To minimize this danger, falcons have an additional secretory gland (a Harderian gland) that produces a viscous (thick) solution especially adapted to moisten the cornea during a stoop.

Figure 10. The function of the small cone or tubercle located inside the nostrils (nares) of true falcons and some other raptor species remains a mystery. It may play a role in olfaction or in modifying the flow of air at high speeds. *(Photo courtesy of Nick Dunlop)*

How Peregrines are able to breathe at such high speeds without bursting or damaging their lungs is puzzling. They have a small cone of bone (a tubercle) that protrudes slightly from the nostril (or nares) that may make breathing easier by serving as a baffle to slow down or diffuse the airflow into the nares during a high-speed stoop. A similarly shaped cone in the center of a jet engine serves that purpose. The problem with this hypothesis is that some other birds have that tubercle, such as the Savanna Hawk *Buteogallus meridionalis* of South America, and they do not dive at such high speeds. And some birds that *are* high-speed flyers, like eagles, do not have a nasal tubercle. No experimental evidence yet exists of the tubercle's function in raptors, although falconer Tom Cade, in his book *Falcons of the World,* suggests that the tubercle may be related to the sense of

Migration Theories

In most instances, the accumulation of empirical knowledge eventually leads to theoretical models that can be tested and refined. To predict the consequences for migrating animals of habitat changes on the earth, we first need to understand migratory phenomena in relation to potential changes in the conditions of feeding sites and breeding grounds that are part of the animals' annual cycle. If some degree of predictability shapes each migrant's annual routine, is a species' vulnerability to habitat change predictable?

In March 2009, a conference on animal migration was convened at the Lorenz Center in Leiden, the Netherlands. Its goal was to move forward the integration of field and experimental ecology with theoretical modeling, and several theoretical approaches to migration were discussed. According to R. McNeill Alexander of the University of Leeds, it is necessary to "identify the key parameters predicting an organism's optimal strategy in the face of environmental variability." Would it be best to stay put, migrate, or go dormant? What are the physical constraints for migrants of varying sizes and modes of locomotion?

To minimize migration time and minimize flight costs, what is the optimal relationship between flight costs and airspeed? We know that both migration and breeding demand great expenditures of energy, but what is the trade-off between migration and breeding? How does the timing of variations in environmental conditions determine when and where reproduction occurs?

Alasdair Houston of the University of Bristol raised the question of the theoretical relationship among speed of migration, predation risk, and arrival time, and Jasper A. Vrugt of Los Alamos National Laboratory and colleagues examined whether migrating passerines (songbirds) "minimize the time spent migrating, or the rate of energy expenditure, or behave according to a compromise."

Migration Theories

Another factor that is apparent but unresolved in any discussion of migration is whether migrants have information about where they are going and how they will get there. Either this is not addressed, or it is apparently assumed that experienced migrants have some awareness of environmental cues with regard to stopover sites and final destinations. In the context of climate change, Italian biologists Nicola Saino and Roberto Ambrosini raised the question of whether migrants are able to "predict future and distant environmental conditions," and if so, which local information would they use for these predictions?

The conclusion of the conference was that several frameworks for theoretical models of animal migration already exist, but that they need to be made more flexible and user-friendly to accommodate additional variables. If these models are made widely available, it is hoped that more scientists will be inspired to focus their work in this direction.

smell or may be sensitive to changes in pressure or temperature in relation to airspeed.

Question 4: How far can a raptor fly?

Answer: Three raptors are generally identified as the raptors that make the longest migrations. Large, crow-sized Peregrine Falcons can be found in most of the world, and many of these populations do not migrate, but the Peregrines that breed in the tundra of Alaska and Canada make one of the longest migrations of any raptor, flying to spend the winter months in central Argentina and Chile. A few of the northernmost migrating Peregrines from Canada and Greenland were electronically tracked, and it was found that their trip took from fifty-six to seventy-two days, traveling a distance of 7,770 to 9,402 miles (about 12,500 to 15,000 kilometers). Although migrating birds typically prefer to fly overland, Peregrines have been seen over the ocean catching and eating seabirds in flight. They have even been seen resting on ships at sea. Although they exhibit strategies to resist being pushed out to sea by unfavorable crosswinds, hawks are sometimes found dead at sea when the winds prevail, indicating that migration can be a costly endeavor.

The medium-sized Swainson's Hawk *B. swainsoni* establishes breeding territories during the spring and summer as far north as Alaska and the southwestern Yukon, down the west coast of the United States through Texas and Mexico, and as far east as Nebraska and Kansas. In the winter, they migrate south to Argentina, Uruguay, and southern Brazil, and smaller populations winter along the Florida and Texas Gulf Coasts. The flight from the northern breeding ground to southern Brazil or Argentina can be as long as 6,200 miles (10,000 kilometers). Each migration can last up to two months, with the birds traveling up to 124 miles (200 kilometers) each day. The small, kestrel-sized Red-footed Falcon *F. vespertinus* migrates some 6,600 miles (11,000 kilometers) each way annually, from eastern Asia to southwestern Asia, and then apparently right across the Indian Ocean to southern Africa.

The raptor that makes the longest migration is the Steppe Buzzard *B. buteo vulpinus.* It is a widely distributed Eurasian buzzard that breeds as far north as the Arctic Circle in eastern Scandinavia, west to the steppes of central Asia into Mongolia, and south to the Northern Caucasus range. Large flocks migrate to spend the months from October to April in a warmer climate, primarily to Africa as far south as the Cape of Good Hope, a distance for some individuals of more than 7,500 miles (12,000 kilometers). They are joined by other species at various points along the migration route, known as the Eurasian–East African flyway.

Edna Gorney of the University of California, Santa Barbara and Yoram Yom-Tov of Tel-Aviv University studied the physical condition of Steppe Buzzards during spring migration as they passed through Israel. They found that adults had significantly

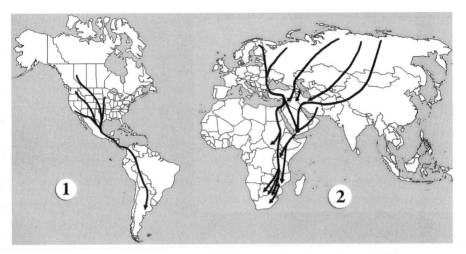

Figure 11. Many raptor species are highly migratory. Shown here are (1) the seasonal movements of the New World Swainson's Hawk *Buteo swainsoni* and (2) the Old World Steppe Buzzard *Buteo buteo vulpinus*. *(Illustration courtesy of George C. West)*

greater fat reserves than immatures, but most of the buzzards they trapped and measured "could not have completed their entire migration using only stored fat: hunting was probably necessary to replenish their energy reserves." Keith Bildstein of the Hawk Mountain Sanctuary reports that raptors have been observed drinking water to supplement the water normally obtained as a result of metabolizing their stored fat. Some long-distance migrants can drink even brackish or salt water, because glands above their eyes that concentrate and excrete excess salt. These glands normally are used to prevent salt buildup when raptors consume dehydrated prey.

Question 5: Do all raptors migrate?

Answer: According to a global directory of raptor migration sites put together by an organization called Raptor Watch, at least 63 percent of the world's raptor species migrate. The increasing use of satellite telemetry to track raptor migrations is providing some more detailed information about the elements that influence the convergence of routes and the hazards of migration.

Migration has a great impact on survival. If migrants' arrival at breeding grounds is delayed, there is an increased probability of breeding failure. A delayed start to a seasonal migration can result in a mismatch between arrival time and food availability at the destination. Delayed avian long-distance migrants, especially those that are insectivorous, may miss the short spring peak when large populations of herbivorous insects forage on young leaves before the production of secondary plant compounds makes the leaves unpalatable.

Millions of birds attempt to cross the barren Sahara Desert as they travel between northern Europe and tropical Africa on their autumn and spring migrations. The Sahara covers more than 3 million square miles (8 million square kilometers), and the shortest distance across the desert is about 930 miles (1,500 kilometers). A successful crossing takes the average raptor more than six days, with essentially no opportunity to refuel. Roine Strandberg of Lund University in Sweden and colleagues used satellite telemetry to record ninety crossings of the Sahara Desert by four species of raptors (Osprey *Pandion haliaetus*, Honey Buzzard *Pernis apivorus*, Marsh Harrier *Circus aeruginosus*, and Eurasian Hobby *F. subbuteo*). Slow travel speeds and aborted crossings yielded a mortality rate of 31 percent per crossing attempt for juveniles, but only 2 percent for more experienced adults. Delayed arrival at their destination was caused by aberrant behavior such as slow speeds, course changes, retreating from the desert, and exhaustion caused by fighting sand storms and severe headwinds.

Question 6: How do raptors find their way during migration?

Answer: There is a fair amount of research about navigation during long-distance seasonal movements, or migrations, but virtually all of it concerns small animals, experimental subjects that are easy to feed and manipulate such as migratory butterflies, bats, and small birds. According to Keith Bildstein of the Hawk Mountain Sanctuary, by the end of the twentieth century, "more than a million raptors had been banded in North America alone." Studies that track raptor migration using satellite or radio telemetry, radar, banding (also called ringing), or marking the birds in some way (clipping a small portion of feathers or color marking by gluing on a dyed feather) have yielded information about their routes, but not about how they navigate. Paul Kerlinger, director of the Cape May Bird Observatory in New Jersey, reviewed the flight strategies of migrating hawks and pointed out some of the reasons for our limited knowledge. Raptors tend to fly at very high altitudes, making it difficult to observe their migration patterns without fitting significant numbers of birds with expensive tracking equipment.

There have been so-called displacement experiments with small raptors. Bildstein described a series of experiments in the 1930s by Rudolf Drost, a German ornithologist, in which he trapped and banded several hundred Eurasian Sparrowhawks and released some of them several hundred kilometers from their home range. He later recaptured some of the displaced birds, as well as some of the control group of birds that had been captured and released without being displaced. The study yielded some useful information about the differences between the behavior of experienced adults and first-time migrants with regard to their ability to redirect their movements so that they could arrive at their normal destination upon being released. The adults were more capable of accomplishing this task, suggesting that experience plays a role in navigation.

Maintaining large raptors in captivity and using them as experimental subjects is a challenging task and is rarely attempted, so discussing how they navigate is speculative. The navigational tools uncovered as a result of research with other long-distance migrants are probably used by migrating raptors, perhaps in some combination, as part of a complex navigational system.

The concepts of an instinctive orientation mechanism and an inherited navigational strategy are frequent subjects of speculation among raptor researchers.

We know that many diurnal migrants use sun-compass navigation (experiments with migrating Monarch butterflies are described in coauthor Carol A. Butler's book *Do Butterflies Bite?*). Many captive birds exhibit migratory restlessness (*Zugunruhe*) at times of the year when they would normally migrate in the wild, and experiments have demonstrated that the length of the day (photoperiod) is what triggers the impulse to migrate. There is some evidence that nocturnal migrants use the stars to orient themselves. Small captive songbirds, Indigo Buntings *Passerina cyanea*, were tested in a planetarium in which U.S. behavioral ecologist Steven Emlen shifted the image of the night sky, with the result that experienced birds adjusted their trajectory to accommodate the shift.

Sensitivity to the earth's magnetic fields has been detected in bats and birds (experimentally exposing them to varied magnetic fields causes them to change the direction of their flight). In at least some birds, structures containing iron-rich particles are present in the mucus skin at the inner edges of the upper beak and in the nasal cavity, according to research by Wolfgang Wiltschko of Goethe Universitat, Germany, and colleagues. When they exposed a control group of migratory Silvereyes *Zosterops lateralis* to a strong magnetic pulse, the birds shifted their migratory flight direction by 90 degrees. When the experimental group had a local anesthetic (Xylocaine) applied to the skin of the upper beak and were then exposed to the same magnetic pulse, they maintained their original heading. This suggests that magnetite receptors in the beak may play a part in navigation for migratory Silvereyes, and perhaps this may be found to apply to raptors as well.

It is clear that, to some degree, experienced migrants use vision, spatial memory, and regional landmarks to find their way or to retrace previous migrations. Bildstein suggests that first-time migrating raptors instinctively orient their movements in a set direction for a certain distance, and that this allows them to flock with migrating adults and learn from the experience of the group, perhaps through some form of cultural transmission. It is possible that flocking makes it easier to locate rising thermal air currents, which are pockets of warm air generated from adjacent features of the landscape that have different heat absorption and reflection characteristics. Riding the thermals enables the birds to spend more time conserving energy by soaring, generally improving their chances of reaching their destination.

Although, like other migratory animals, they use stored fat as their primary energy source, large raptors making very long migratory flights need to eat and rest along the way. They conserve energy by soaring and gliding on thermal air currents, and observation of their well-established flyways or migration routes shows that they choose to migrate over land rather than water, giving them access to prey and safe places to rest. Narrowing land masses in southern Florida and Mexico become bottlenecks as North American migrating birds converge and concentrate over the land, attracting birding enthusiasts to these locales where it is possible to observe thousands of migrating raptors every day (see figure 11 showing migration corridors). Eliat, at the southernmost tip of Israel, is another ideal observatory in the spring when raptors migrating north to their breeding grounds converge to rest after flying 1,245 miles (2,000 kilometers) across the Sahara and Sinai deserts from their winter homes in southern Africa.

Question 7: How do raptors hunt?

Answer: Different birds of prey use different hunting methods. Falcons and some eagles fly high in the air and dive or stoop to attack their prey in midair. Falcons bite the neck of their prey whether or not it is already dead. Buteos such as Rough-legged Hawks *B. lagopus* and Swainson's Hawks perch on tree limbs and drop onto ground-based prey animals. The Secretary Bird (see color plate C) walks along to hunt for snakes and other small prey, and the Osprey and some eagles dive into the water to catch fish, sometimes briefly submerging.

Most owls have special feathers that allow them to fly silently, so they fly low, usually at night, and stealthily approach and pounce on their unsuspecting prey. Many species of nocturnal owls begin hunting when darkness falls. Raptor expert Jemima Parry-Jones cautions motorists to take extra care when the clocks change in the fall due to daylight savings time. Cars kill many owls at this time of the year as people begin driving home from work after dark. It takes the owls a week or two to adjust to the additional amount of traffic on the roads when they come out to feed.

The Crane Hawk *Geranospiza caerulescens* (see color plate D) has a very special hunting technique using its double-jointed legs. It climbs a tree and inserts a leg into a hole in the tree, bending it upward and downward to grab any nesting birds or sleeping bats inside. The African Harrier Hawk *Polyboroides typus* hunts in a similar manner. Vultures (see color plate B) generally eat carrion, inserting their head into the animal's body to feed on its innards. The Bearded Vulture *Gypaetus barbatus* flies into the air with a tortoise grasped in its bill or feet and then drops the tortoise onto rocky ground to crack the shell and expose the meat. It also flies up holding large, bare animal bones and drops them onto rocks to splinter them so it can swallow the shards of bone. The Egyptian Vulture *Neophron percnopterus* picks up small rocks and drops them on an egg until the egg breaks. They also pick up small eggs and drop them on the ground to crack them open.

Question 8: Are birds of prey social or loners?

Answer: Raptors are basically loners, defending their territories in a variety of ways. John and Frank Craighead of the U.S. Fish and Wildlife Service studied nesting hawks and found that territories vary in size according to species, and that breeding pairs have home ranges where they hunt that may overlap with territories of adjacent pairs. As a general rule, the smaller the bird, the smaller its range, but the more likely it is to defend its territory by attacking intruders. Some species have favorite hunting places that they defend, and migratory birds may adopt a temporary territory in their winter quarters. The pair bond may loosen during nonbreeding periods, especially in migratory pairs, and the birds may temporarily roost alone or even in flocks in their winter habitat.

Question 9: How do raptors communicate?

Answer: Owls vocalize with low-pitched sounds commonly described as "hoo hoo" or "hooting," but not all owls hoot. Barn owls *Tyto alba* screech and purr and make a variety of other noises. Nocturnal species (night flyers) make more sounds than do species that are diurnal (day flyers), and they also tend to have better hearing, since hearing and communicating with sound go hand in hand. Owls use sound to claim a territory, to attract a mate, to warn, and to communicate with a mate and offspring.

Some raptors also communicate by flying in dramatic patterns in courtship or nuptial displays, and there is some thought that the obvious mating activity of some breeding pairs is a form of territorial display, communicating to other birds that the territory is already occupied without having to engage in a physical confrontation (see chapter 4, question 1: How do birds of prey attract a mate? and question 3: How do raptors mate?).

Question 10: Are raptors always aggressive?

Answer: Aggression is strongly instinctive in raptors, as it is in other predatory animals such as lions, crocodiles, and sharks. To survive, raptors must single-mindedly pursue, kill, or disable their victims without being injured in the process. This is usually a vicious business, but the ferocity of the raptor helps minimize the possibility that the prey will escape or resist. If a raptor were less aggressive and struggled with prey, an injury from a defensive bite or a jabbing beak would probably cost the raptor its life. The injured predator would become prey, or it would die of starvation because it was unable to hunt.

Figure 12. Bald Eagles *Haliaeetus leucocephalus* with talons locked in midair. The immature bird on the right is attempting to pirate the fish in the adult's beak. *(Photo courtesy of David Hemmings)*

The aggressive nature of these birds often has deadly consequences to nestling raptors. In general, female raptors are larger and more aggressive than males, and a hungry female nestling, impatient for her parents to return with food, may kill and eat her brothers. This happens especially in certain hyper-aggressive hawks such as the Goshawk *A. gentilis* and the Cooper's Hawk *A. cooperii*. Although predatory birds are by necessity and instinct aggressive, adult birds are actually quite gentle and affectionate while brooding and feeding their young. The parents of newly hatched hawks, falcons, and eagles tear tiny bits of flesh from captured prey and gently coax the babies to take the morsels from their beaks, a scene that shows a raptor at its gentlest moment.

Question 11: Are raptors dangerous to people?

Answer: In general, raptors are not a threat to people. Despite their reputation as fierce and ruthless hunters, raptors do not fly around grabbing people with their talons. (The old tales of huge eagles carrying off human infants have never been substantiated, although theoretically it is possible that some of the larger eagle species that prey on monkeys could carry a small baby.) For the most part, it is the nature of wild raptors to avoid conflict with humans and other animals that they do not perceive as prey, because the risk of injury and death is too great. But hand-reared raptors, particularly large species, can be quite aggressive toward strangers and even toward their handlers, and they can inflict injuries if not handled carefully.

Figure 13. After an exciting chase, a falconer catches some rest while his Harris's Hawk *Parabuteo unicinctus* feeds on the jackrabbit it has captured. (*Photo courtesy of Robert Leporati*)

In circumstances in which wild raptors feel threatened by a human, however, they can in fact pose a danger. The sharp bill or talons of an injured raptor may be dangerous to a person who is attempting to subdue it for treatment (see chapter 8, question 4: What should I do if I find an injured or dead raptor?). Some raptor species defending their nest, eggs, or young from a human intruder can inflict serious injury. The Kenyan ornithologist Leslie Brown reported that the large and powerful African Crowned Hawk-Eagle *Stephanoaetus coronatus* is not shy about flying at and striking a person who dares to approach its nest too closely. Heavy leather jackets and hard hats like those worn by workers at construction sites is standard protection for researchers, banders, or falconers who climb to the nests of Northern Goshawks, as this species will repeatedly dive-bomb and strike an intruder while uttering loud, defensive calls.

Aggression within a species can vary from region to region, most often in relation to differences in nest defense and to the degree of human pressure to which they have been exposed. Bald Eagles *Haliaeetus leucocephalus* in the Aleutian Islands, where they have seldom seen humans or firearms, have been known to come relatively close to observers and occasionally have attacked them. In continental North America, they tend to maintain a much greater distance from observers and are much more wary. Falconers are sometimes injured by their birds, almost always incidentally in the course of the training process. Sometimes a trained raptor's talon will slip into a seam in the glove and pierce the falconer's flesh. Or the bird may bite or scratch when it is feeding on the glove and the falconer adjusts the food near the bird's feet or beak with the unprotected hand. Occasionally a falcon swoops at and strikes a falconer who is trying to train the bird by swinging a lure (a pair of dried pigeon, pheasant, or duck wings tied together with a tidbit of meat attached to simulate live prey). For more explanation of the training process, see chapter 7, question 7: How are raptors trained?

FOUR

Raptor Reproduction

Question 1: How do birds of prey attract a mate?

Answer: Raptors attract a mate in the breeding season by calling loudly and by flying in dramatic displays. Most accipiters perch singly and call (vocalize) at regular intervals. Single raptors of many genera, usually males, may soar and circle over their breeding area to attract attention, and they may also indulge in undulating displays, repeatedly swooping down almost to the ground from a considerable height and then rising sharply. Some soaring and screaming may occur in groups of several birds among a few species, but usually these displays are solitary endeavors. Some species engage in nuptial (sexual) displays by soaring together and touching or grasping one another's feet as they fly.

Because many species of owls are nocturnal, less is known about their courtship behavior other than that they call repeatedly to advertise their eligibility. Diurnal owls engage in flying courtship displays, sometimes clapping their wings loudly as they fly. Male Barn Owls *Tyto alba* screech and hover in midair before a female, and like other raptors, try to attract her to the nesting site by offering gifts of food.

Question 2: At what time of year do birds of prey mate?

Answer: The precise time of year that a species of raptors begins looking for mates varies, but in general, the desire to breed

Figure 14. A pair of Ospreys *Pandion haliaetus* copulate on a dead tree near their nest. Mated pairs of raptors may copulate many times a day, especially at the beginning of the breeding season. *(Photo courtesy of Darryl Luck)*

is triggered by incremental increases in the length of the day, which coincides with warming weather and an increase in the populations of prey animals. The additional sunlight stimulates the birds' pituitary gland, sending hormones into circulation that cause the male's testes and the female's ovary (many birds only have one that develops fully) to increase in size. The instinct to engage in species-specific courtship rituals is also stimulated as part of this process.

Great Horned Owls *Bubo virginianus* are among the earliest birds in North America to begin looking for a mate each year. They start courting in late December and can be sitting on eggs by late January or early February. Breeding and raising offspring are energy intensive for both males and females, but because these owls are generalists that eat a wide variety of

Figure 15. These Bald Eagles *Haliaeetus leucocephalus* are about to lock talons and tumble toward the ground at high speed, engaging in one of the elaborate and breathtaking aerial courtship displays common among raptors. *(Photo courtesy of Gary Woodburn)*

animals, they can find prey and be well nourished even in cold weather. Insect-eating owls, in comparison, do not breed until the weather is warmer and insects are more abundant.

The Red-tailed Hawk *Buteo jamaicensis*, whose diet consists mainly of small rodents, has a more typical breeding cycle. These hawks start breeding in the early spring, and females lay up to three eggs six to eight weeks after mating. Both sexes brood the eggs for about a month until the young hatch in the late spring.

Most raptors breed every year, but if their favored prey is not abundant and available one year, they may not look for a mate. This is especially true of larger birds that feed on small mammals with strong population fluctuations. Some very large eagles, like the Harpy Eagle *Harpia harpyja*, breed only every two or three years, but not for lack of prey. Their young take approximately six months to leave the nest and almost another year before they are able to feed themselves independently, so the parents are occupied with each brood for almost two years. Smaller species with a shorter lifespan are under more pressure to breed every year.

Question 3: How do raptors mate?

Answer: Mating typically follows complex courtship displays that may involve calling, flying, and feeding the female (described in this chapter, question 1: How do birds of prey attract a mate?). In an established pair, mating may occur without this preamble when the female indicates she is receptive. Male and female birds each have only one opening, a cloaca, which is used for copulation and for excreting wastes. The male's sperm, produced in the testes, passes to his cloaca, where it is stored until copulation. The female's cloaca leads to her ovaries. When she is ready to mate, the female crouches down and spreads her tail, moving it to one side, and the male settles on her back, curling up his talons to avoid injuring her, and maneuvers his tail away so that their cloacae can make contact. The sperm is transferred to the female in a few seconds in what is called a cloacal kiss.

The female stores the sperm for at least a week (in some species for much longer) and then fertilizes each ovum with the stored sperm as it moves from the ovary into the oviduct. The avian ovary contains follicles at different stages of development, and they develop in sequence to a rapid final growth stage, which enables the female to fertilize and lay one egg after another as each ovum matures and is released. She lays a clutch of several eggs, one at a time, with a day or two between eggs.

During the breeding season when hormone levels are elevated, the male bird's testes become considerably larger than normal in order to produce sperm, with the left testis usually

Figure 16. Position-
ing their tails to the
side, Peregrine Falcons
Falco peregrinus make
cloacal contact during
copulation. The male
positions his toes so
as to avoid scraping
the female's back with
his talons. *(Illustration
courtesy of R. David
Digby)*

more enlarged. Sara Calhim of Queen's University in Kingston,
Ontario, and Tim Birkhead of the University of Sheffield in the
United Kingdom demonstrated that if one testis is impaired or
fails to develop, the other serves as a backup and compensates
for the reduced function. Hatchling female birds possess two
ovaries, but in many species only the left one develops and becomes functional, enlarging during breeding season to produce
the ova. Aaron Peterson of Hope College in Michigan and colleagues found that in Tree Swallows *Tachycineta bicolor,* males
initiated cloacal contact from the left side three times more
frequently than from the right, suggesting that the directional
bias may be an adaptation to the location of the ovary. This interesting angle on copulation in birds may also apply to raptors
and other birds, although we have not seen any data to either
support or refute this.

Figure 17. During the breeding season, a female raptor's ovary looks similar to this ovary of a domestic hen *Gallus gallus* bred for continuous egg production. Egg follicles are in various stages of development, although a raptor will produce far fewer eggs. *(Photo courtesy of Alan Johnson)*

Fewer than 3 percent of birds have an intromittent organ (penis), but some male raptors have a small cloacal protuberance during the breeding season because the seminal vesicles lie near the skin surface at the cloaca. In a few species of birds, including ostriches *Struthio camelus,* cassowaries, kiwis, swans, geese, and ducks, the males do possess a retractable penis. Since waterfowl sometimes copulate in water, one hypothesis is that the penis may ensure that the sperm is deposited far enough inside the female so that it is not flushed away by the water.

Except in ducks and geese, order Anseriformes, where the male has an intromittent organ, aggressive or forced copulation in birds is rare. When mating involves the cloacal kiss, the awkward location of the cloacae on the ventral (lower) part of the male and female bodies requires that both male and female cooperate to accomplish successful sperm transfer. Especially in falcons and accipiters, the male is very cautious during mating and keeps his talons clenched when mounting the female. Females of these two groups are significantly larger than males, and they are clearly in charge of this process: any inadvertent scrape to the female might result in her attacking or injuring the male. Further, females can resist fertilization by using their powerful cloacal muscles to eject sperm as they do waste material, according to Patricia Gowaty of the University of Georgia and her colleague Nancy Buschhaus.

Although sperm can be transferred from the male to the female cloaca in an instant, and a single copulation seems to be sufficient to fertilize an entire clutch, female birds typically copulate repeatedly with their mate or copulate with multiple partners. A hypothesis about why this occurs, proposed by Michael Lombardo of Grand Valley State University in Michigan and colleagues, is that the female may receive a cloacal inoculation of beneficially sexually transmitted microbes that either can be therapeutic or can offer her protection against future exposure to pathogens due to the excretory function of the cloaca and the potential exposure to intestinal microbes incorporated into the male's ejaculate. James Briskie and Robert Montgomerie of Queen's University, Ontario, suggest that the *absence* of an intromittent organ requires explanation, and they hypothesize that one reason most male birds do not have a penis could be to minimize the risk of contracting pathogens and parasites during copulation, due to the dual function of the cloaca.

Question 4: Are raptors monogamous?

Answer: Most solitary (noncolonial) species of raptors appear to be monogamous, although research that sequences the DNA of nestlings has revealed that extrapair copulation does occur. Many raptors, especially larger eagles like the Golden *Aquila chrysaetos* and the African Crowned Hawk-Eagle *Stephanoaetus coronatus* remain mated for life in a territory that they will fight to keep for themselves, using the same nest year after year. Some smaller and medium-sized raptors also mate for life, and nonmigratory pairs can be seen traveling together searching for prey during the nonbreeding season. In the American Kestrel *Falco sparverius*, according to Leslie Brown and Dean Amadon, "sexual activity may begin some weeks before nesting; . . . two or more males have been observed mating with the same female without hostility. . . . When nesting begins, the species is monogamous."

Copulation patterns have been studied for several species of diurnal raptors, in part because their copulations are conspicuous and frequent. According to J. J. Negro and J. M. Grande of the Spanish Council for Research: "The majority of raptorial birds for which precise estimates are available copulate more than 100 times per clutch," with the American Kestrel reportedly copulating up to 690 times per season. The researchers suggest that all this obvious copulation is a type of territorial display or signaling. It may serve to defend the territory from intruders by clearly demonstrating that there is a pair in residence, while avoiding the risk of a physical confrontation. Sperm transfer could be wasted if the purpose of the mating display is primarily to deter potential intruders, and in fact, the researchers have observed that many of these "copulations" occur without cloacal contact, providing support for their display hypothesis. They predict that "colonial raptors will have higher copulation rates than solitary ones because the number of potential signal receivers is higher, and that nocturnal raptors will not tend to use copulation as a territorial display because it cannot be observed in darkness."

Mate guarding and frequent copulation, two main methods of assuring paternity in birds, are the focus of considerable research based on hours of intensive observation. In mate guarding, the male closely monitors his mate and tries to physically prevent her from engaging in extrapair copulations. With most raptors, close mate guarding is not possible because males must frequently leave the nest to hunt to provide food for the female during the breeding period, when she stays at the nest for several weeks. Frequent copulation within the pair may allow the male to dilute sperm deposited by rival males and thereby decrease the risk of copulation with a nonmate (cuckoldry). Frequent copulation may also serve to maintain the pair bond.

Tim Birkhead and Catherine Lessels of Sheffield University in the United Kingdom contend that intrapair copulation to attain paternity assurance in the Osprey *Pandion haliaetus* is more frequent in colonial species in which the risk of extrapair copulation is high because of nearby males. Supporting this position, Per Widén and Malena Richardson of Karlstad University in Sweden found that copulation occurred more frequently in Ospreys in higher density breeding areas. Anders Pape Møller of the University of Aarhus in Denmark found the same pattern in the Northern Goshawk *Accipiter gentilis,* as did Robert Simmons of the University of the Witwatersrand in Johannesburg in African Marsh Harriers *Circus ranivorus.* Earlier research by Simmons, and later by Erkki Korpimaki at the University of Turku in Finland, contended that the frequency of extrapair copulation among raptors is relatively low and unrelated to the species copulation frequency. On the other hand, Francois Mougeot of the Centre d'Etudes Biologiques de Chizé in France analyzed the literature in 2004 and found support for the hypothesis that the intensity of sperm competition increases with breeding density, and that male raptors rely on frequent copulation to ensure paternity.

Mougeot and colleagues used male and female decoys to manipulate the risk of extrapair copulation in raptors, and they noted that extrapair *fertilization* rarely occurred in raptors, suggesting that their technique of guarding the female to try to assure paternity is effective. Yu-Cheng Hsu and colleagues of National Taiwan University studied the behavior of the Lanyu Scops Owl *Otus elegans botelensis* and reported that "reproductive success can be estimated accurately by simply counting young in nests." They found only two cases of "parentage mismatch" among 108 broods. David Arsenault of the University of Nevada and colleagues reported in 2002 that the socially monogamous Flammulated Owls *O. flammeolus* that nest in groups were genetically monogamous despite their high breeding density. However, some subsequent research using DNA paternity tests on nestlings found that some of them were not fathered by the male partner.

In a five-year observation period, Katrine Eldegard and Geir Sonerud of Norwegian University of Life Sciences fitted radio transmitters to Tengmalm's Owls *Aegolius funereus* parents and offspring to explore the strategy they employed to care for their young. They observed that 70 percent of the females deserted their offspring, leaving the males to provide care. When they experimentally supplemented the food available in the habitat, the researchers found that a higher proportion of females deserted, and they deserted when the offspring were younger. In general, fewer offspring in deserted broods survived, and the researchers postulated that the benefit of deserting may be that the female can mate again.

Question 5: Do raptors of one species mate with other species?

Answer: Raptors of one species generally do not mate with other species in the wild for several reasons. William Eberhard, an expert on sexual selection, hypothesized that "courtship behavior probably serves as a preliminary filter of possible mates, and 'incorrect' cross-specific pairings resulting in genitalic contact probably occur less often in species with more complex preliminary behavior." Many raptors use dramatic flight patterns and calls to attract mates (see this chapter, question 1: How do birds of prey attract a mate?), and one species' vocalizations and behaviors are unlikely to be attractive to a potential mate of a different species—they may even stimulate alarm. Size and color differences among different genera and species also inhibit mating because the birds do not recognize one another as potential mates. These morphological (physical) differences are reflections of very different chromosomal patterns, indicating probable genetic incompatibility that also plays a part in their lack of interest in crossbreeding. Geographical barriers such as mountain ranges, rivers, and forests often create reproductively isolated populations of raptors that physically do not have the option to interbreed because they are not in proximity to one another.

Sometimes falcons of different species court, mate, and copulate on their own in captivity and—rarely—in the wild. In the early 1970s in Ireland, the first hybrid falcons were produced unexpectedly when falconers Ronald Stevens and John Morris housed a Peregrine Falcon *F. peregrinus* and a very large Saker Falcon *F. cherrug* together. If the species are closely related, interspecific coupling may result in the production of fertile eggs and the hatching of hybrid young. Lynn Oliphant of the University of Saskatchewan reported a cross in the wild between a Peregrine Falcon and a Prairie Falcon *F. mexicanus* that resulted in two healthy male offspring. But these are unusual examples. Naturally occurring hybrids are rare, and most hybrids are infertile. Usually, falcon breeders create hybrids by artificial insemination (see this chapter, question 6: How is artificial insemination practiced with raptors?).

It is illegal in the United States to release hybrids into the wild, although some scientists believe that hybrids are morphologically or behaviorally different enough from wild populations that they are unlikely to breed with wild birds. Scientists are concerned that resources could be strained if hybrids were released. There is also concern that the genetic consequences of hybrids breeding with wild birds could cause an imbalance or even contribute to the extinction of a species. This may be an alarmist viewpoint, because genetic dilution of the hybrid parent occurs rapidly, after only three to four generations. In addition, species may be much more closely related than was earlier thought. DNA analysis by Franziska Nittinger from the Museum of Natural History Vienna and colleagues has shown that there is a very close genetic similarity among some species in the genus *Falco*. One result of these studies places Saker Falcons and Gyrfalcons in the same clade, meaning they share features inherited from a common ancestor. We know that in captivity these two species often produce fertile offspring when hybridized.

Question 6: How is artificial insemination practiced with raptors?

Answer: The breeding of raptors of the same species in captivity has been pivotal in saving them from extinction. The numbers of some species plummeted in the wild due to the indiscriminate use of the pesticide DDT and the loss of suitable habitat caused by human expansion. Renz Waller in Germany in the 1940s pioneered a captive breeding program, producing a few Peregrines. His work was followed by that of Frank Beebe in British Columbia (1967), Heinz Meng in New Paltz, New York, in the 1970s, Tom Cade and colleagues at Cornell University in the 1970s and 1980s, and Richard Fyfe and colleagues in Alberta from the 1970s through 1996. A whole host of raptor breeders became involved to contribute to the release of captive bred Peregrines (not hybrids) to the wild after the environmental problems had been mitigated. Species once thought doomed have made remarkable recoveries.

Raptors are bred in captivity for falconry with the goal of producing a "quality eyass (young) raptor," according to Peter Gill, a raptor breeder for over twenty years. The bird, almost always a hybrid, should have the best physical, behavioral, and genetic qualities to insure its future breeding success. Gill explains that the commercial breeding of raptors must be tailored to the market. Perhaps an Arabian falconer, for example, wants a Gyrfalcon/Saker Falcon hybrid to go hawking for Houbara Bustard *Chlamydotis undulate,* a game bird, or a British falconer wants a Gyrfalcon hybrid or Peregrine to hunt Red Grouse *Lagopus scoticus.* Permits or licenses may be required, depending on where the breeding takes place.

To inseminate a female artificially, seminal fluid containing sperm cells from a male raptor is introduced into the oviduct of a female who has laid her first egg. The theory is that introduced sperm will unite with viable egg cells high in her oviduct, fertilizing them before egg-white proteins and the beginnings of a shell surround the ovum. The tapered, smooth tip of a plastic or glass pipette containing semen is gently placed into the end of the oviduct via the cloaca. Years ago, one blew gently into the open end of the pipette to force the fluid through the tube into the oviduct, but modern pipettes are fitted with a plastic handle and air is pushed through by turning a wheel on the handle with your thumb. Throughout this process, the female falcon is "cast," meaning she is hooded and held breast down on a pillow or other soft cushion with her wings closed (usually with a towel or piece of cloth wrapped around her). Inseminating the female is fairly easy and straightforward. Getting seminal fluid from a male falcon is the tricky part.

Retrieving semen from a male falcon is accomplished by one of two methods. Male falcons imprinted on humans have a desire to copulate with them during the breeding season. Taking advantage of this bonding, a researcher can wear a special falcon-semen-collecting helmet to collect semen from imprinted male falcons. This device is made of sturdy smooth plastic and looks like a telephone lineworker's hard hat, with a deep continuous moat instead of a brim around the whole of the circumference. A block of wood covered with a piece of carpet is secured to the top of the helmet and serves as a perch. Imprinted male falcons fly onto the perch and copulate, sending seminal fluid dribbling down the sides of the plastic and into the moat at the brim. It is then an easy matter to retrieve the semen from the helmet for use in artificial insemination.

The second method of collecting semen from a male falcon is known as "stripping." A male falcon is held gently or cast by one person, while another lightly strokes his back and sides near the region of his kidneys and testes. If the bird is in breeding condition, seminal fluid begins to dribble down the vas deferens, the tubes that connect the testes to the cloaca. At this point, a hand-automated suction pipette is inserted into the cloaca and vas deferens, and, with a push of the thumb on the pipette wheel, a milliliter or so of clear amber-colored falcon semen can be sucked into the pipette tube. This sample may be used immediately or frozen for later use or for shipping to another facility.

Before the advent of the hand-automated pipette, someone on the team of falcon strippers who served as the "sucker" had the unenviable job of sucking on the open end of the pipette to extract semen from the male falcon. Coauthor Peter Capainolo had some experience performing this procedure as a college undergraduate. Despite keeping an eye out for the rapid movement of seminal fluid up the tube, he occasionally learned the hard way that while falcon semen looks like a nice lager, it tastes rather bitter, possibly because the white, foamy head of this brew is highly concentrated uric acid.

Question 7: Do all birds of prey make nests?

Answer: Boreal Owls *A. funereus,* Northern Saw-whet Owls *A. acadicus,* and other small owls usually lay their eggs in tree cavities that have previously served as nests for other birds. Since they do not build their own nests, many of these species readily accept the safe, enclosed space a nest box offers (see chapter 8, question 6: Of what value are raptors to the environment?). Winter storms in the frigid habitat of Fish-Owls (genus *Ketupa)* frequently snap off parts of trees, and once the exposed surfaces begin to decay, the owls use the weathered cavities for nests (see figure 18). Barn Owls nest in buildings, in mine shafts, and under bridges, and they tend to use the same site for years, of-

Figure 18. Blakiston's Fish Owls *Ketupa blakistoni* lay only one egg (or, rarely, two) every other year (the second egg in this nest never hatched). Due to the harsh conditions of their far north habitat, fledglings are up to a year and a half old before they disperse. *(Photo courtesy of Jonathan C. Slaght)*

ten returning to it with a new mate if the previous one dies. The Short-eared Owl *Asio flammeus,* a grassland denizen, and the Arctic Snowy Owl *Bubo (=Nyctea) scandiacus,* build simple nests on the ground. Burrowing Owls *Athene cunicularia* nest in abandoned underground burrows that formerly housed ground squirrels or other small animals.

New World vultures and the genus *Falco* and related genera also do not build their own nests. They may take over an abandoned nest, or they may lay their eggs in a hollow tree or in a protected spot on a cliff ledge or rocky outcrop. Peregrine Falcons are native to a wide variety of open habitats, but they usually nest in an isolated spot that is near good hunting grounds. On rare occasions they have been observed nesting in abandoned tree nests. The Canadian Peregrine Foundation reports that Peregrines have been seen "especially in Scandinavia" nesting "on the ground in bogs where they are as safe from predators as they would be on a cliff. Most nests, regardless of where they are located, consist simply of a shallow depression scraped out by the adults; no nest materials are added to these "scrapes." The Rough-legged Hawk *B. lagopus* usually chooses a site under a rocky overhang to protect its nest from snow, but sometimes they nest right on the ground, perhaps at the brink of a bluff. Some large raptors tend to return to an old nest and repair it year after year. Some hawks build nests out of sticks and use them for only one season, leaving the nests to be taken over by owls.

Raptors usually prefer to nest in a high location to minimize predation. Peregrines are reported to prefer protected ledges from about 50 feet to as high as 200 feet above ground (15 to 60 meters), preferably with a southern exposure. Brown and Amadon point out that in the plains of Tibet and Turkestan where cliffs are rare and desirable nesting sites are scarce, one craggy spot may be occupied by breeding birds of several raptor species that ordinarily would keep their distance from one another.

Before choosing a nest site, the male of most raptor species must find a breeding territory big enough to support enough prey to feed his future family. In addition to a promising site for a nest, if he is a nocturnal owl it must include a safe roosting spot for daytime sleeping. The supply of food in the territory is a factor that limits its population, and the size of a breeding territory varies greatly among individuals of the same species depending on the availability of food. Another influential factor is the territorial aggressiveness of the species that may already inhabit the territory—sometimes a raptor that finds a desirable territory will be able to evict the current tenant. The age at which a raptor is mature and ready to search for a territory varies among the species. Most owls, for example, are mature after about one year and begin seeking their own territory in their second year.

Peregrines' scrapes may be as close as half a mile (less than a kilometer) from each other, but if food is not abundant, a pair will defend a territory that is much larger. In the northern United States where there are dense populations of Peregrine Falcons, Clayton White, formerly of Brigham Young University, and colleagues found that nests averaged between 1.5 and 2.5 miles (3.3 and 5.6 kilometers) apart. Home territories ranged from about 77 to almost 600 square miles (200 to 1,500 square kilometers), and males and females hunted more than 2 miles (5 kilometers) from their nest site or territory. According to naturalist Helen Roney Sattler, Great Horned Owls need a territory of only 1,000 acres (4 square kilometers) to accommodate their requirements. Flammulated Owls need only about 7.5 acres (just over 30 square meters), and Great Grey Owls *Strix nebulosa* need only a few hundred feet. Wu Yi-Qun at Lanzhou University in China and colleagues found that the minimum breeding density of Long-legged Buzzards *B. rufinus* in their study area in northwest China was 0.19 breeding pairs per 100 square kilometers. James Woodford of the Wisconsin Department of Natural Resources and colleagues surveyed Red-shouldered Hawks *B. lineatus* in northern Wisconsin, and found that their reproductive success was relatively low in a study area where there were very few nests (a nest density ranging from 0.10 to 0.16 nests per square kilometer). The authors concluded that in order to maintain this breeding population, it appeared to be critical to have greater nest density.

Observations of a Barn Owl Nest

The Barn Owl *Tyto alba* is a bird of prey essential to healthy ecosystems because they eat large numbers of rodents. It is the most widely distributed species of owl, found almost everywhere in the world except at the poles and in the deserts. These owls, which fly silently and low to the ground, feeding on rodents and other small vertebrates, make up the family Tytonidae; all other owls are placed in the family Strigidae. Barn Owls' unique facial discs inspire some of their local names, such as heart-shaped or monkey-faced owl. These owls are fairly large, with longer legs than other owls, dark eyes, and no ear tufts. A golden-buff color dominates the upper body, and the face and underparts are usually white in males and more rufous in females, with some individuals exhibiting dark spotting. As with most raptors, the female is larger than her mate. Large birds may reach lengths up to 20 inches (about 50 centimeters), but despite their size, they are quite light and fly with a mothlike agility.

My introduction to Barn Owls up close began as I sat on the back porch of a friend's home in Brookhaven on Long Island's South shore one spring, looking out over a huge field of mixed weeds. My friend mentioned the regular nighttime appearance of a large owl that cruised over the fields and always returned toward the nearby evergreens, often making many trips back and forth during the night. His detailed description made the bird sound more and more like a Barn Owl, and I was excited by the prospect of a family of Barn Owls in the area because I had never observed the species closely, and I knew they were becoming less common.

We decided to head for the pines with a flashlight to see what we could discover. As we approached the trees, a large owl exploded from the cover and scolded us with a short, rasping scream. The bird was close enough that I could definitely identify it as a Barn Owl. The scolding bird continued its distress call and was joined by another, smaller owl—presumably

(continued)

the male. From beyond the tree line came a call I had never heard, hissing and high-pitched, like a red-hot horseshoe dipped in a bucket of cold water. Obviously more than one animal was making the noise. We pushed our way through the vines and ground cover and came upon an old water tower some 30 feet high. The hissing was coming from inside, and somewhere above us were the young of our Barn Owls.

The next day I went back with my camera and inquired at the house nearby to see if the owner might know something about the owls. The woman who lived there said that owls had raised young in that tower for the past thirty-four years and had been monitored for many years by a local ornithologist. She kindly gave me permission to observe them at my leisure.

I found a door cut into the water tank to permit my predecessor to gain access, and the climb up was made simple by a ladder welded into the metal framework of the tower. The water tank was open to the sky, and its floor was about five inches thick with a carpet of castings composed entirely of small rodents called voles, genus *Microtus*. When I reached the top of the ladder, I saw the female squatting in her nest, eyes closed, in front of five owlets. She and the three older siblings were swaying slowly from left to right, all the while making the now-familiar hissing sounds. The youngest owlet appeared to be no more than one day old, and the oldest seemed to be about one and a half weeks. The female continued swaying but never attacked, not even a few days later when I picked up the young for banding. I never saw the male during the day, but I assumed he was not far off in a tree somewhere, observing my actions.

I thought surely the youngest owlet would become breakfast for his siblings, but all five fledged just fine. They grew quickly and soon looked more alike with their downy plumage. By the middle of July, the white down had been replaced by gray. The feather shafts of young birds are soft and filled

Observations of a Barn Owl Nest

with blood, and by the end of the month their feather shafts had become hard and mostly hollow; the wing and tail feathers were now "hard penned." Coverts and contour feathers grew out completely by the end of August, just before fledging. About a week prior to their fledging, I had to leave, but my friend informed me that the parents fed the young in the conifers surrounding the tower for a week before the five dispersed. I regretted having missed the last stages of fledging, and I was determined to catch the entire breeding cycle the following summer. Now that I had banded the owlets, I hoped to discover if later generations returned to nest in the place where they had hatched.

—Peter Capainolo

In most species that build a nest, the male chooses the general territory but the female chooses the specific location for the nest. She frequently does most of the construction, although the male may supply some of the construction material. The time spent preparing a nest by birds of the same species ranges from a few days to several months, proceeding in a more leisurely fashion in a tropical climate and going more quickly where the weather is seasonal and cold weather looms. In the Arctic, there is no time to waste, and nests must be prepared quickly if the young are going to survive the winter.

Question 8: Do raptor parents share nesting duties?

Answer: In general, raptor parents do share nesting duties since most raptors mate for life. They share in building or refurbishing the nest, and they take turns incubating the eggs. The female does some of the hunting but leaves the bulk of that responsibility to the male. She occasionally leaves the nest to defecate, perhaps cough up a pellet or two, and stretch. Once the young raptors grow some feathers and can stand and walk around in the nest, they instinctively void over the edge of the nest, helping to some degree to minimize its accumulating mess. Some Eastern Screech Owls *Megascops asio* catch small blind snakes and place them in the nest with the young. The snakes eat insects that are attracted to the droppings and food scraps, and this helps to keep the nest clean; otherwise, raptor parents do not bother cleaning the nest. When the young have grown enough feathers to keep themselves warm, the female begins to leave the nest more frequently, flying with the male to hunt for food.

Question 9: How many eggs do various species of raptors lay?

Answer: Most diurnal raptors lay from one to three eggs, with two or three being most common. Intervals of two or three days are usual between depositing each egg. The average clutch for an owl may be as many as five eggs, although there is a wide range across species. Most birds begin incubating only when all the eggs have been laid over a period of several days, which results in all the young hatching at approximately the same time. But owls that begin laying eggs in cold weather must begin to incubate immediately so that the first egg does not freeze, with the result that their eggs hatch one by one over a more prolonged period.

In the late nineteenth century, there was a huge upsurge of interest in natural history collections by scientists and ordinary citizens. Victorian homes were commonly decorated with collections of stuffed birds, and oology, the study and collection of eggs blown clean of their contents, was all the rage. Magazines like *Ornithologist and Oölogist* made it possible for collectors to buy and sell eggs around the world. One collector claimed to own "a hundred and eighty Peregrine Falcon clutches comprising more than seven hundred eggs, out of a collection totaling twenty thousand of many species," according to Scott Weidensaul in his history of U.S. birding. Although it could become excessive, the passion for collecting led to a great deal of the knowledge about birds that we can now simply query on the Internet.

As a general rule, smaller birds of prey lay larger clutches, and clutches laid at higher, cooler latitudes are larger than clutches laid by birds of the same species in a hotter climate. Some species have larger clutches when ample prey are available and smaller clutches when prey are scarce. Snowy Owls have been known to lay as many as fourteen eggs in years when there is plenty of food, and none at all in years when food is in short supply. Although there is a relationship between the abundance or scarcity of food and the size of the clutch, the availability of prey also affects the number of young that are successfully reared in each nest (see this chapter, question 12: How does a raptor female feed her nestlings?).

Question 10: What do raptor eggs look like?

Answer: Owls lay white eggs that are more or less round in shape. Other raptor eggs are more oval, and some are speckled or softly colored. Large Great Horned Owls lay eggs about the size of a chicken egg; smaller Pygmy Owls *Glaucidium californicum* lay eggs that are half that size. Raptor eggs vary a great deal, but compared to other birds, their eggs tend to be large for the size of the parent bird. The color of the inside of the shells ranges from white to greenish, and these colors are useful in identifying the species of an egg found without any associated cues.

Scott Weidensaul tells a wonderful story about a young field scientist, Major Charles E. Bendire, who, as a cavalry officer riding in the wilds of Arizona, collected bird eggs and shipped them to Spencer Fullerton Baird in Washington, D.C., for the collection Baird was amassing for the Smithsonian's National Museum of Natural History. One day in 1872, Bendire spotted an all-black raptor that was probably a Zone-tailed Hawk *B. albonotatus,* a rare local species. Anxious to collect its eggs, he left his shotgun with his horse and climbed 40 feet up a tree to reach the bird's nest. From that height, he spotted a group of

Apache crouched about 80 yards away. He descended carefully, trying not to reveal to the Apaches that he had seen them, and he popped the egg he had retrieved into his mouth to keep it safe, "and a rather uncomfortably large mouthful it was, too." He rode the 5 miles back to camp with the egg in his mouth, "expecting an attack at any moment, cradling the egg against the jolts and jars of the ride, his jaw muscles swelling, trying to breathe, trying not to gag. And then, safely to camp, trying to remove the precious egg without breaking it, his breath coming in labored gasps."

Question 11: What is the likelihood of survival for a raptor nestling?

Answer: Many birds die in the first year of life, even before leaving the nest. This is a function of natural selection, since more birds are produced than can possibly survive. A conservative estimate is that 20 to 40 percent of nestlings probably die as a result of starvation, siblicide (being killed by a sibling), or predation. Cynthia Berger estimates that 50 to 70 percent of owls die before they reach one year old. The time immediately after fledging (leaving the nest for the first time) is the time of highest mortality. From fledging until birds reach maturity and acquire their adult plumage, estimated mortality is as high as 90 percent for some species due to starvation, predation, disease, and accidents (see also this chapter, question 13: How long does it take before raptor young fledge?, and chapter 3, question 5: Do all raptors migrate?).

Question 12: How does a raptor female feed her nestlings?

Answer: When first hatched, nestling raptors are too weak to eat. Female hummingbirds and some other birds force-feed their nestlings by placing regurgitated food directly into their mouths, but raptor young must quickly learn to take small bits of food from the hooked beak of the parent. If the youngster is too weak to make this effort, or if older nest mates outcompete it, it will not survive. It is not unusual for the oldest nestling to attack a younger sibling, and since the eldest may be considerably larger, the younger one can be killed or injured. It may fall out of the nest in an effort to escape, or it may simply starve to death. When the parent arrives with food, it feeds the oldest nestling first because it is the most aggressive in soliciting food from the parent. If prey is scarce, the youngest usually starves, because there is not enough food left once the eldest is satisfied, and feedings occur less frequently because the parents are making fewer kills.

Once the nestlings' tarsus (legs) and bill have developed sufficiently, the young birds are able to tear up the prey the parents bring to the nest. They attack the food and instinctively adopt a threatening posture over it, called "mantling," which is the parents' cue to spend less time in the nest.

Question 13: How long does it take before raptor young fledge?

Answer: Young birds of prey are quite helpless (altricial) when first hatched, and for larger species, a relatively long period precedes trying to fly or leaving the nest (fledging). Small falcons and Sparrowhawks A. nisus fledge in three to four weeks, as do small Pygmy Owls, but nestlings of eagles and other large birds like Great Horned Owls can take up to ten weeks or longer to fledge. Offspring of one of the largest raptors, the California Condor Gymnogyps californianus, fledge after five or six months but are not completely independent until they are two years old.

Until they fledge, the young birds eat voraciously and their weight can increase fiftyfold. Fledging tends to occur more quickly in temperate and Arctic regions than it does in the tropics, because the impending cold weather puts pressure on the young to be strong enough to leave the nest and migrate to avoid starving to death. In the Arctic when it is light even at night, parents feed their young twenty-four hours a day.

The nestlings at first stay close to the mother. She broods them almost constantly, keeping them warm and protecting them under her feathers until their second coat of down appears, at which point she may brood them only at night since they are more able to withstand the cold. At this stage they can stand and walk around the nest, and soon feathers start appearing through the down in a predictable pattern. A person knowledgeable about the developmental details of a species can estimate a young bird's age by observing which feathers are visible. Now the young birds are more active, practicing skills they will need as adults: picking up nesting material, calling, flapping their wings, and being more attentive to their environment. Before they are ready to fly away from the crowded nest and live independently, they spend at least a week or two as "branchers," sitting on nearby branches, sometimes falling to the ground, but still being fed and guarded by their parents. At the end of the fledging period they may be in or near the nest, and they simply fly away one day, motivated at least in part by hunger. Their flying ability quickly reaches an expert level, and then they are truly on their own.

FIVE

Dangers and Defenses

Question 1: Which animals prey on raptors?

Answer: Raptors sometimes fight over prey, and the weaker bird may be left dead or seriously injured. Smaller raptors, such as Eurasian Sparrowhawks *Accipiter nisus,* may be killed by larger raptors like Northern Goshawks *A. gentilis.* The Eurasian Eagle Owl *Bubo bubo* preys on Peregrine Falcons *Falco peregrinus,* both juvenile and adult, especially during their coinciding breeding seasons. The most vulnerable birds are the nestlings. The various species of small Harriers, genus *Circus,* nest on the ground and fly low to hunt small mammals and birds, making them vulnerable to predation by foxes, coyotes, and other mammals. In Africa raptors may fall prey to baboons or leopards.

Question 2: How do prey animals defend themselves against raptors?

Answer: Most prey animals try to hide or flee when they become aware of a bird of prey nearby. Many raptors prey on birds, and some species of birds have developed special defenses that may seem paradoxical. In "mobbing," a group of prey animals harasses an animal that is usually their predator. The fitness benefit or evolutionary advantage of any behavior must equal or exceed its costs for it to be maintained, and mobbing meets this test. Noisily mobbing a raptor signals the predator that it has been detected, removing any advantage that it might have gained by surprise. Mobbing gives the prey animals close-up information about the age and condition of the predator, helps them assess the level of risk it poses, and may discourage an attack by demonstrating the prey's alertness and vitality. It also informs others in the group and in the habitat about the potential threat. Ryo Ito and Akira Mori of Kyoto University found that spiny-tailed iguanas *Oplurus cuvieri cuvieri* in Madagascar reacted with increased vigilance to mobbing calls made by the Madagascar Paradise Flycatcher *Terpsiphone mutate*. Both species are preyed on by raptors but otherwise have no relationship (they are not predator-prey and do not compete for resources).

Michael Griesser of Uppsala University in Sweden exposed groups of Siberian Jays *Perisoreus infaustus* to mounted specimens of three hawk and three owl species that posed varying known risks to Jays in that habitat. When the predator mount was exposed near a group of Jays, the Jays immediately interrupted foraging and mobbed the mount, emitting a greater number of calls when mobbing the mount of a more dangerous predator. Griesser found that the sounds they made were "predator-category-specific," signaling with hawk-specific or owl-specific call types when mobbing that bird. This demonstrates that mobbing calls can encode complex information to warn kin and group members about the level of risk from potential predators. Jessica Yorzinski and Gail Patricelli of the University of California, Davis, had similar results in their study of ten species of passerine birds that they exposed to a mount of a Great Horned Owl. Each bird was placed in a small cage surrounded by eight microphones. The birds' vocalizations were more focused when they were directed at the perceived predator and more diffuse when the caged bird was communicating with nearby birds of the same species.

In some instances mobbing escalates into attacking and chasing the predator, possibly to move the predator away from the vicinity of the prey and, if it is breeding season, away from their mate and offspring. A fledgling American Kestrel *F. sparverius* in Brooklyn, New York, was rescued from a flock of pigeons that was chasing and pecking at the young bird (the rescuers named the bird Alice Cooper because of its facial markings). Because this was an immature bird, the full meaning of this event is not clear, but it may be an example of mobbing that had become an attack.

Specific take-off noises made by wing structures in certain pigeons have recently been found to serve as alarm sounds. A recent study by Mae Hingee and Robert Magrath of the Australian National University found that Crested Pigeons *Ocyphaps lophotes* have modified flight feathers that produce a distinct whistle in alarmed flight. The researchers were able to signal alarm to a resting flock by replicating the sound.

Question 3: How do raptors defend themselves?

Answer: Raptors can be very aggressive in defending themselves from predators that are equal or larger in size, using their strong talons to "foot" the attacker, but they also have a range of less risky defensive behaviors like posturing, flight displays, and chasing. Clint Boal of the University of Arizona described two physical encounters between Cooper's Hawks in the city of Tucson during their breeding season that may exemplify their defensive style. In one exchange, "an adult male chased an intruding adult male out of the nest stand and pursued it for over 550 yards (500 meters), neither bird ever flying more than 21 yards (20 meters) above the ground. On three occasions during the chase the hawks faced each other while hovering, and repeatedly made stabbing strikes at each other with their feet. Contact was made several times, but neither male maintained a hold on the other." In a second situation, a "subadult" female was circling above the nest, and the adult female on the nest "made several passing strikes at the intruder, which rolled and extended its talons toward the aggressor. The subadult was struck solidly at least three times before leaving the area." Other intrusions Boal observed did not elicit aggressive responses from the adult on the nest. In each of four instances, the nesting adult appeared to ignore a potential intruder that perched nearby, and the intruders eventually flew away after observing the nest for several minutes.

The Turkey Vulture *Cathartes aura* may vomit defensively to repel an intruder. It expels some of the partially digested contents of its stomach in the direction of the intruder, repelling it with the smell and also causing discomfort if the vomit lands on the intruder's skin or in its eyes. If the vulture has undigested food in its crop from a recent kill and needs to flee from a predator, it may create a diversion by disgorging the morsel of food at the feet of the aggressor. This serves a dual purpose: it lightens the vulture so it can fly away more easily, and it tempts the predator with a meal that is more appealing than the fleeing vulture.

Vultures have an interesting defense against bacteria and other microscopic dangers. A vulture's head is bare of large feathers. When feeding on a dead animal, vultures typically insert their head into the decomposing carcass to reach the meat, and head feathers would be likely to accumulate organic material and bacteria from the carrion. After feeding, the birds typically sun themselves, using the heat of the sun to dry out any debris that may have adhered to their skin so that they can shake it off. They also urinate on their legs, and the acidic urine probably kills any bacteria remaining on their legs from the latest meal. The evaporation of the urine may serve the additional purpose of dissipating some excess body heat in hot weather.

An owl in a tree defends itself by adopting a concealing posture, squeezing its body together so that it looks thinner and taller, even manipulating its facial feathers to hide its eyes and beak. Owls' plumage provides excellent camouflage, and if you have ever gone into a wooded area looking for owls, you know they usually are well hidden on a high branch. The clues to their presence are at the base of the tree—the telltale white "wash" from their urine and their regurgitated pellets.

Question 4: What illnesses occur in wild raptors?

Answer: If an animal in the wild shows signs of weakness, it is likely to be killed and eaten in short order. Even a predator like a raptor, if it is debilitated, will quickly become prey. If a raptor looks sick, it is probably already near death, so diagnosing and treating an apparent illness is quite challenging. Pathogenic infectious microorganisms, allergic reactions, toxins and genetic disorders are all possible causes of disease in captive and wild raptors.

Because all warm-blooded mammals can get rabies, there is suspicion that birds that eat mammals may transmit the rabies virus through their feces, although there is no evidence that this has happened. L. M. Shannon of the University of California, Davis and colleagues examined fifty-three newly captive birds of prey for rabies antibodies and reported that "no significant antibody titers were detected" in any of the birds. According to Jim Parks and Julie Collier, Massachusetts-based raptor rehabilitators: "Disease is an infrequent problem, since raptors are free of many of the illnesses such as rabies and sarcoptic mange (a skin disease caused by mites) that plague wild mammals." The Ohio Department of Natural Resources concurs with this view: "When an animal is sick with rabies, the virus is shed in the saliva. It is then spread to other animals or people when the virus-laden saliva gets into a wound or mucous membrane . . . usually through a bite. . . . Birds and reptiles do not carry rabies."

Question 5: What injuries are common among wild raptors?

Answer: A questionnaire was sent to 65 of the 250 identified raptor rehabilitation facilities in the United States in 1994 by Patrick Redig and Gary Duke of the Raptor Center at the University of Minnesota They asked each facility to identify the main causes of injury in the birds brought to them. Of the thirty-two rehabilitators who responded, over half said the top cause of injury was vehicle collisions. Next most common was trauma, such as flying into windows and power lines, followed by orphaned birds, shooting, and toxicity. A study in 2007 conducted by Amy Pauli and colleagues at the University of Minnesota College of Veterinary Medicine found that in the rehabilitation centers she queried, the largest category of injury in the raptors they admitted was eye injuries due to trauma (ranging from 14 to 28 percent).

Impact injuries are frequent. Because of changes in the landscape due to development, raptors may become confused and disoriented and sometimes fly into windows, buildings, wind turbines, and structural towers. Cars hit raptors hunting in roadside grassy areas that attract small prey animals. Airplanes can be deadly to raptors and other birds, and "bird strikes" by large birds can damage a plane and even disable its engine. From 2000 to 2008, the Federal Aviation Administration logged almost five hundred collisions of birds with airplanes in the United States.

Another source of injury to raptors is illegal hunting. Under the Migratory Bird Treaty Act, permits to hunt or capture raptors are granted only for research, rehabilitation, falconry, and limited Native American ceremonial purposes. To discourage poaching, even the possession of a feather without the proper permit is a punishable offense (see chapter 9, question 5: Are any raptors endangered?).

Cats and other ground-dwelling animals are a danger to raptors, especially to fledglings that are strong enough to leave the nest but not quite able to fly. Known as "branchers," these young birds may flutter out of the nest and end up on the ground at the base of the tree where their nest is located, unable to fly back up to the nest. The parents will continue to feed them on the ground until they can fly, but they can easily become prey if their parents are not nearby to chase away predators.

In many areas in the world, overhead power lines are strung on millions of poles that are the tallest points in the landscape. Raptors often use these structures as nest sites and as preferred vantage points from which to hunt. But many of these installations are not properly insulated or not insulated at all. The risk of electrocution is minimal for small birds, which usually make contact with only one energized component. But a raptor with a wide wingspan may touch two energized components or an energized part and a ground wire at the same time, completing a circuit and exposing itself to electrocution or serious injury.

HawkWatch International's Raptor Electrocution Reduction Program surveyed more than thirteen thousand power line structures in Utah from July to November 2001, and found 196 electrocuted birds at their bases, most of them raptors. Although the problem is most acute in treeless areas, urban-dwelling raptors can be at risk from power lines as well. Because of that danger, a study conducted in Tucson, Arizona, by James Dwyer and R. William Mannan recommended that "all potentially lethal poles within three hundred meters of the nests of urban-nesting raptors be retrofitted through the addition of insulation, or through increased spacing between conductors."

In the landmark Moon Lake case, at least 170 carcasses of electrocuted raptors, about three-quarters of them Golden Eagles, were found under power lines in Western Colorado and Eastern Utah owned and maintained by Moon Lake Electric Association. The U.S. Fish and Wildlife Service warned the group to make improvements to their equipment and ultimately subjected them to criminal prosecution. In August 1999, Moon Lake pled guilty; the group paid $100,000 in fines and restitution, their CEO was put on probation for three years, and they agreed to create an Avian Protection Plan (APP) with a timetable to guide them in retrofitting their equipment. An APP establishes an internal remedial process that is activated when an electrocuted bird is found at the base of a company's power pole. It addresses safe construction standards for new lines, develops retrofitting standards for existing equipment, and often includes a proactive risk assessment that involves inspecting existing installations.

Although retrofitting is voluntary, the government's victory in the Moon Lake case set a precedent, and penalties can be severe for companies whose installations kill eagles. Misdemeanor company fines can be as high as $200,000, and individual officers can face up to one year in jail; more severe, felony-level penalties may apply in some situations. Rick Harness, a wildlife biologist at EDM International, developed the Avian Protection Plan (APP) required in the Moon Lake settlement. Colorado is the only state where every electric cooperative and investor-owned utility responsible for power lines has an APP, and EDM

Figure 19. Bald Eagle *Haliaeetus leucocephalus* on a power line covered with a raptor hood to protect large birds from accidental electrocution. Eagles have a wider wingspan than the 60-inch industry standard between power lines and can easily be injured or killed if they contact both wires. *(Photo courtesy of Charlene Burge)*

has developed more than sixty APPs for entities from Florida to Alaska. The company is currently working on a plan for installations in the Mongolian steppes, where the power lines have been electrocuting Saker Falcons *F. cherrug.*

Massive commercial wind-powered generators (wind turbines) are becoming common sources of environmentally friendly power, but as evidenced by dead birds and bats found at the base of the turbines, they cause some injury and death. Power output increases with the range of the rotor blade's sweep, and a turbine blade can be as much as 130 feet long (40 meters), able to turn more than 220 miles per hour (360 kilometers per hour). Wind speed increases with altitude, so taller generator towers capture more energy, generate more electricity, and when they are idle, appeal more to raptors as nest sites and vantage points from which to hunt. Turbines are powered up or down depending on changes in the wind, and raptors perching on an idle turbine's

The Hunter and the Hunted

Ardent falconers, particularly those who have been around for a while, fairly overflow with tales of the hunt. One of my favorites takes place on a cloudy day in mid-December 1981, when I was an undergraduate at the State University of New York at New Paltz. It was late in the afternoon as I drove west out of town with my female passage Merlin *Falco columbaris* to steal the last half hour of daylight to exercise her with the lure.

Since I had captured and trained her earlier that year, she had taken a decent number of European Starlings *Sturnus vulgaris*. I was heading for an area we called the flying fields, so named because Heinz Meng, my mentor and professor, and other falconers as well have flown birds there for many years. The area treats the visitor to a spectacular view of huge fields, brown in winter, that heave and roll westward, bisected by a gravel road, and in the distance the magnificent Shawangunk cliffs rising sharply from the deep forest. There are some hedgerows scattered on this landscape and deciduous plots surround it. A huge, ancient, dead hardwood tree with high, twisted branches that stands alone in the middle of the fields has served many raptors—trained, passage, or winter residents—as a vantage point. Today a lone Kestrel *F. sparverius* sat at the top of the big tree. These smallest of North American falcons are less compact and heavy than their Merlin cousins. They hunt mainly mice and insects. This one was a male, his bright blue wings a cold grey in the dimming light.

Snow threatened as I took the Merlin up from the cadge, a type of portable, padded perch that has been used for centuries to transport raptors. Outside the car, the frozen ground crackled and my fingers numbed in the few seconds it took to remove the Merlin's hood. I cast the Merlin off my light glove.

There was no wind, and the Merlin stooped at the lure in tight circles. She began to range farther from me, trying to

(continued)

come in lower and faster, a change in strategy designed to increase her chances of taking the lure, as well as making it clear that she could choose to leave.

On one pass she didn't "throw up" as expected, meaning she didn't shoot straight into the air and down to the lure again. Instead, she continued on toward the tree where the Kestrel perched. Since she was low to the ground, I knew she had no designs on the Kestrel. Apparently startled by her approach, the Kestrel took flight, dipping low in front of her, and the stimulus was too much. She was after him. They headed toward my left, both chattering constantly, the Kestrel sounding similar but higher pitched than the Merlin. I stood transfixed and useless with the lure line dangling from my ungloved, frozen right hand. As they drew closer to me, the gap between them closed and they looked like one long-necked, four-winged mutant bird.

Suddenly, a passage male Cooper's Hawk *Accipiter cooperii* zipped from behind my right shoulder so close that I could hear the whoosh of his wings. In an instant he was behind my Merlin, the trio now engaged in some bizarre bird-eat-bird struggle. My mind raced as I watched in fascinated horror. I did not want the Merlin to catch the Kestrel because they are not legal quarry, but even less did I want my carefully trained Merlin to provide a meal for the sleek hawk. Realizing that the hunter was now the hunted, the Merlin swerved downward as the Kestrel continued, frazzled but unharmed, into the distance.

Although falcons are generally fast, my Merlin was tired and the Cooper's Hawk was persistent and capable. He stalled several times and grabbed at my bird, missing by inches while I held my breath. About 200 yards from me, the Merlin crashed deep into the safety of a hedgerow at the field's edge. Seconds before the impact with the hedge, the Cooper's Hawk fanned his long tail and veered vertically, alighting on the hedge just above the spot where the Merlin sat hunkered

The Hunter and the Hunted

down and fluffed. Still intent on eating her, he craned and swayed his head jerkily, his bulging yellow eyes searching for an opening.

Before he could make a move, I became a lunatic, swearing and running toward him hurling pebbles and clods of dirt. He looked up then down again, not wanting to give up. As I got closer, instinct prevailed and he lifted off and drifted easily into a tree near the main road. He was still watching as if he might get a second chance.

I tried to fathom what had just transpired as I made for the frightened Merlin, who warily looked around after stepping onto my fist. On the dark drive home, sitting on the cadge and hooded, she seemed content. I took longer to settle down than she did, and I realized I had witnessed something unique. In what must have been only a few minutes, I had watched a rare interaction of three species of raptors; two birds were wild, and although my Merlin was trained, in a very real sense she still was wild, as is true of all passage raptors.

As falconers we are focused on the interactions between predator and prey, but once in a while we get a glimpse of the relationship among the predators themselves—one reason falconry continues to be a source of unexpected thrills and excitement.

—Peter Capainolo

structures are at risk of being flushed off when the blades begin rotating. The force generated by the blades can cause wind shear, forcing a bird to the ground if it is in front of the generator. In addition, whirling vortices at the ends of the blades can tumble a bird into the structure or onto the ground.

Migrating raptors take advantage of updrafts on the sides of mountains or ridges, and these slopes are favorite places to locate generators, so collisions are inevitable. Other concerns are the noise level of the turbines, which may interfere with a

Figure 20. In midflight, this Merlin *Falco columbarius* has flipped on its back to snatch a dragonfly out of the air, although its preferred prey is small birds. *(Photo courtesy of Richard Ettlinger)*

Figure 21. Smallest of the American falcons, the American Kestrel *Falco sparverius* displays all the physical characteristics of the group, such as long pointed wings and a prominent malar stripe or moustache. *(Photo courtesy of Richard Ettlinger)*

raptor's hearing, and motion smear, confusing images of the rapidly turning blades, which may interfere with its vision. These types of disruption may disorient raptors as they try to fly around the turbine, resulting in broken wings and legs or decapitation and death (see color plate H).

James Castle of Sonoma State University used night-vision devices to observe and report on the presence of nocturnal owls and diurnal hawks and kestrels at a commercial wind turbine facility in California. He found that, in addition to the owls, some diurnal raptors were active around the turbines at night, especially when there was bright moonlight, favorable ambient temperatures, and ample vegetation in the surrounding area to attract prey. Additional research is identifying potential problems so that assessments of the potential effect on wildlife can guide conservation-minded micrositing (determining the best specific location for each turbine).

Question 6: What other dangers do raptors face as a result of development and population growth?

Answer: One example of the danger population growth can pose to raptors concerns Burrowing Owls *Athene cunicularia* in the U.S. Southwest. These owls, quite small and widely distributed, make their homes in the abandoned burrows of prairie dogs, ground squirrels, and other small mammals. Due to rapid development in the Sun Belt, one area in New Mexico went from having twenty owl burrows to just three. The Migratory Bird Treaty Act protects Burrowing Owl nests, but *only* while eggs or fledglings are in the nest. The nests can be legally destroyed once the birds have migrated and the nests are empty (see color plate G).

The California Burrowing Owl Consortium developed mitigation guidelines in 1993 that call for passive relocation of the owls where their presence conflicts with development interests. The technique used successfully in northern California was described in 1995 by Lynne Trulio of San Jose State University. In early spring, before the breeding season starts, a colony is carefully studied. Rather than capturing the birds and transporting them to a new site, a nearby location is chosen that has suitable ground for breeding habitat. The new site is enhanced by adding burrows and perches that are appropriate for the species. Once the owls become accustomed to the changes and show interest in the new location, they are excluded from entering the old burrows that are going to be razed. Howard Clark and David Plumpton of H. T. Harvey and Associates developed a simple one-way trapdoor that is placed over the burrow entrance for this purpose, and in most cases if everything has been done carefully, the owls will move to the new site in the course of a few nights. Between 1997 and 2004, several hundred owls were successfully passively relocated in this manner from more than twenty projects in California.

The New Mexico Burrowing Owl Working Group was formed in 2001 in response to the population decline in their area. They are observing owl colonies and compiling results from fourteen counties in New Mexico, and trying to educate landowners, citizens, and builders about ways to protect the owls.

Northern Spotted Owls *Strix occidentalis caurina* in Oregon are vanishing due to unanticipated competition from another species of owl. Twenty years ago, the northern subspecies of spotted owl was at the center of environmentalists' efforts to save what remained of the old-growth forests that were the birds' habitat. In 1990, as a result of these efforts, the owls were included in the Endangered Species Act, and a federal court ruling in 1991 closed a great deal of the twenty-four million acres of federal land in the Northwest to logging in order to protect the habitat for the owls. By the end of the twentieth century, timber harvesting on federal land had declined 90 percent compared to levels prior to the legislation.

Barred Owls *S. varia* are somewhat larger and more aggressive medium-sized birds that sometimes breed with Spotted Owls. At one time they were found only in eastern North America, but since the federal land was closed to logging they have expanded into the Pacific Northwest and invaded the habitat of the Northern Spotted Owl. Spotted Owls are selective eaters, but

Barred Owls are generalists, eating almost anything—including Spotted Owls. In a 1990 survey, Barred Owls in a forest west of Corvallis, Oregon, occupied less than 2 percent of Spotted Owl sites, but by 2009, they were found nesting in an estimated 50 percent of Spotted Owl sites, clearly able to outcompete the local species for space and food. New recovery plans are being formulated to try to preserve the habitat for the Spotted Owl.

Question 7: Does lead in the environment affect raptors?

Answer: Lead poisoning from rifle bullets, shotgun pellets, and fishing sinkers is a recognized threat to birds, and many countries and localities have already banned these items. Lead poisoning in birds has been documented to cause physical and behavioral changes that weaken the bird, impair its ability to fly and forage, and make it vulnerable to predators. Raptors normally reject contaminated food as unpalatable, but if food is scarce, they are more likely to risk feeding on carcasses of animals that have been shot by hunters. They may actually consume the lead fragments or shotgun pellets, or they may ingest tissues of the carrion that have been contaminated by the lead.

Fish and water birds inadvertently eat lead shot and small sinkers, mistaking them for the pebbles they normally eat to help digest their food (see chapter 2, question 4: How do raptors digest their food?). Ospreys *Pandion haliaetus* and eagles capture fish and waterbirds that may have consumed lead and feed them to their nestlings, passing along the toxins.

The number of California Condors *Gymnogyps californianus* fell significantly due to lead poisoning, in addition to habitat destruction and poaching. In 1982, only 23 birds could be found in the wild. The San Diego Zoo began a captive breeding program that was expanded by the U.S. Fish and Wildlife Service in 1987 to include the Los Angeles Zoo. All the wild condors were captured and placed in breeding programs in the zoos, and starting in 1991, some of the birds have been carefully reintroduced into the wild. As of April 2009, according to a report by the San Diego Zoo, 322 condors were counted, of which 172 were living in the wild.

Lead contamination can be avoided by using environmentally friendly, lead-free ammunition and fishing weights made of tin, bismuth, copper, steel, and tungsten-nickel alloy that presumably do not harm wildlife, although we have not seen any research reports confirming the safety of hunting with "organic" bullets. Old lead gear should never be discarded into water or on land— it should be taken to a hazardous materials collection site.

Question 8: Has DDT affected birds of prey?

Answer: During the 1950s and 1960s, aerial and ground spraying of DDT (dichloro-diphenyl-trichloroethane) was a widely used method of pest control. Swiss chemist Paul Hermann Müller received a Nobel Prize in 1948 for his role in developing this extremely effective pesticide. DDT is persistent, water resistant, and highly toxic to aquatic life, and it is readily absorbed in soil. Indiscriminate spraying was common until the 1970s, without considering its impact on wildlife and human health. DDT's breakdown products and metabolites last for years, accumulating in all levels of the food web, with the highest concentrations in apex predators such as raptors. Raptors are at the apex or highest point of the web because they have virtually no predators of their own. They consume contaminated animals that have fed on toxic plants and insects, and thus accumulate the highest concentrations of the pesticide.

Two examples of birds of prey pushed almost to extinction by DDT's toxicity are a subspecies of the Peregrine Falcon *F. p. anatum* and the Osprey. Formerly known as the Duck Hawk, this Peregrine Falcon subspecies fed on ducks and other birds that had become toxic from consuming DDT-laden fish. DDT accumulated in the fatty tissue of the falcons and caused a metabolic disturbance that resulted in the shells of their eggs being thin and excessively fragile, easily crushed by a light touch from the attending parents or by the pressure of the female's attempt to incubate. In some cases, eggs were laid without shells. The species' ability to reproduce was essentially destroyed, and it became extinct on the East Coast of the United States because of the overuse of DDT.

The U.S. Fish and Wildlife Service created a recovery plan for the peregrine in 1977, revised in 1984, which focused on monitoring nesting activities and contaminant levels. Wildlife organizations assembled recovery teams to develop a strategy, and Peregrine Falcons were bred in captivity by various individuals and organizations with good results (see chapter 4, question 6: How is artificial insemination practiced with raptors?). In just one example of successful captive breeding, more than 250 birds were released from a program in Virginia. The banning of DDT in 1972 made it possible for released birds to breed successfully in the wild, and as a result of the programs in North America, breeding pairs of Peregrines can now be found in the United States, Canada, Mexico, and the United Kingdom, many of them in urban areas. In 2005, for example, a survey found eighteen pairs nesting in New York City (see chapter 8, question 5: What attracts raptors to live in cities?). The anatum subspecies is now only found in western North America and Central America. Peregrine Falcons are still listed in many states as endangered, but they are no longer federally listed.

In the 1950s when DDT use was prevalent, 90 percent of breeding pairs of Ospreys, which feed mainly on fish, disappeared from the Atlantic coast between New York City and Boston. The ban on DDT along with new regulations have helped the Osprey population rebound. In 1984, a hundred nesting pairs were counted in a New Jersey survey, and in 2006, the population in that area had increased to four hundred pairs. Ospreys are federally protected, but only active nests (often located on power poles or other structures) require federal permits to be removed. In Florida, nests without eggs or flightless young are deemed inactive, and they may be removed with a permit issued by the Florida Fish and Wildlife Conservation Commission. But under Florida law, if an inactive nest is removed, a replacement nest structure must be installed in the immediate vicinity of the old nest if at all possible.

Question 9: Which raptors are particularly vulnerable to environmental toxins?

Answer: All raptors are doubly susceptible to environmental toxins because, in addition to direct exposure from aerosol or ground spraying, they consume live or dead prey that itself has been contaminated by consuming the toxins. Vultures are particularly susceptible, however, because they feed primarily on dead animals, so they consume some carrion that contain levels of pathogens and toxins high enough to have caused their deaths. They also are vulnerable to the relatively high doses of veterinary drugs that accumulate in the tissues of large domestic ungulates such as cattle and sheep. Further, vultures eat large amounts and they are anatomically equipped to swallow large food items. A fully distended crop can hold an amount of food equal to as much as 20 percent of the bird's body weight, enough to maintain the bird for three or four days or to kill it with a lethal dose of toxins.

The populations of some of the eight species of vultures on the Indian subcontinent, genus *Gyps,* collapsed in the 1990s and continue to decline. The Indian White-backed Vulture *Gyps bengalensis* and Indian Long-billed Vulture *G. indicus* have suffered 80 to 90 percent declines, and other vultures in that genus have been affected, but less profoundly. Surveys indicate that there has been an extremely high mortality rate for adult birds due to kidney damage, which was traced to diclofenac, a nonsteroidal anti-inflammatory drug that has been widely used in veterinary medicine for treating livestock. The manufacture of diclofenac for veterinary use was banned but its sale is still legal, so it will be some time before the drug is removed from the vultures' food supply. Captive breeding of three vulture species has begun to provide birds for future reintroduction.

In Europe, the Mediterranean region is the core of the species' range, and Spain hosts the great majority of breeding pairs. Livestock carrion is their primary diet, and vulture nestlings in central Spain have been affected by the residue of multiple antibiotics in the carrion that is fed to them by their parents. The affected species are Griffon Vultures *G. fulvus*, Cinereous Vultures *Aegypius monachus*, and Egyptian Vultures *Neophron percnopterus*. The parents feed primarily on stabled livestock carrion that has been treated with antibiotics such as semisynthetic penicillins and enrofloxacin. Jesus Lemus, Guillermo Blanco, and colleagues from the Museo Nacional de Ciencias Naturales in Madrid reported that exposed nestlings demonstrated bacterial antibiotic resistance, making them more vulnerable to bacteria found in carrion and to normal bacterial flora such as *Escherichia coli* in their own system. Their immune systems were found to be depressed when compared with nestlings that did not have circulating antibiotics in their system.

Colonies of Griffon Vultures nest on cliffs in seventeen countries in the Mediterranean area, potentially foraging at great distances from their colony when prey is scarce locally. Because they are scavengers, they have played a role in keeping some of the area's residential neighborhoods clean. The earliest major population decline occurred during anti-predator campaigns of the 1950s and 1960s. Farmers concerned about rabies transmission used poisoned bait, usually containing strychnine, to kill jackals and bats that were considered predators of domestic animals (in reality, none of the bat species in the area feed on anything except nectar and insects). Vultures died in a matter of hours after feeding on the poisoned carrion.

In the 1980s and 1990s, wolves and feral dogs had become problematic predators of domestic animals, and poisoning became popular again, further contributing to the depletion of the vulture population. There were struggles in Israel over pasturing areas, and farmers poisoned each other's cows, contributing to the extinction of vultures in the Galilee area and a dramatic drop in nesting pairs in the Golan Heights.

New settlements have continued to displace nesting vultures in the Mediterranean area, uninsulated power lines electrocute a number of the birds annually, and hikers and photo-hungry tourists disturb birds, causing them to abandon optimal nest sites. In Sardinia, for example, only two vulture colonies remain.

Crete has some small colonies that have a low occupancy rate. The most detailed population data about the Griffon Vulture seems to be available from Israel, collected and reported in 2000 by Reuven Yosef of the International Birding and Research Center in Eliat, Israel, and Ofer Bahat of the Israel Institute of Technology. They estimated 1,000 breeding pairs in Israel in the mid-nineteenth century that by the 1950s had all but disappeared. In the late 1970s and early 1980s, about 80 pairs were reported. In the late 1980s there was an increase to 130 pairs, but as of 1998 they report that only 45 pairs had successfully reared young. Subsequent data from the Israel Nature and Parks Authority reports 400 individuals in 2001, diminishing to 220 individuals in 2009.

Illustrated books for children about saving the birds, written by Ran Levy-Yamamori and available in Hebrew and Arabic, have been distributed at no charge to every family in regions where the poisonings occurred, part of an effort to rehabilitate the vultures and reestablish their populations. The Israel Nature and Parks Authority developed a mitigation plan in 2008 that involves marking (tagging) some birds to study their movements. In addition, some eggs are removed from nests at the Gamla Nature Reserve, a practice that encourages the breeding couple to lay another egg. The removed eggs are incubated artificially at the Biblical Zoo in Jerusalem, and when the birds hatch they are kept in captivity until age three, when they will be released if things go according to plan. The hatchlings are fed using a special glove to prevent imprinting (see figure 22 of a vulture puppet used for feeding, and chapter 6, question 4: What is imprinting?).

Recently there have been alarming reports that cattle ranchers in Kenya are deliberately poisoning lions and hyenas that threaten their herds. It is suspected that they have been using Furadan (carbofuran), an odorless and tasteless toxin manufactured in the United States as an insecticide. Eagles and vultures that prey on the dead lions and hyenas have been poisoned as well, and this chemical may be responsible for what appears to be a significant reduction in the populations of vultures and

Anti-predator Hunting

In 2009, a controversial anti-predator campaign was initiated in the United States. Its goal was to limit a predator population without creating the cascading effect on the environment that has resulted in the past from the widespread use of poisons (see chapter 5, question 9: Which raptors are particularly vulnerable to environmental toxins?). The federal government idealistically inserted a predator that unbalanced the local ecosystem, providing a complex illustration of the delicate balance between reintroducing an endangered species and its impact on predator-prey relationships.

Wolves were once common in North America, but they were largely killed off in the Northern Rockies by the mid-1930s. They were listed as endangered in the United States (except in Alaska) in 1974, and in 1995, a federal wildlife program began reintroducing gray wolves *Canis lupus* in Idaho, Wyoming, and Montana. By the end of 2008, the wolf population had reached five times the goal set in the reintroduction plan. The northern Rockies gray wolf population was considered "fully recovered" and was taken off the endangered species list in May 2009. In Idaho, for example, about 850 gray wolves were counted at the end of 2008.

As a consequence of the increase in the wolf population, there were frequent reports of wolves killing cattle and sheep and depleting the deer and elk populations that are favored by the area's big-game hunters. Hundreds of wolves were legally killed to protect livestock, but the need for appropriate future management of the wolf population became a controversial issue. Many local people criticized the federal government's action. In this area of the country, hunting is an important activity, and some hunters saw the large wolf population as unwanted competition.

Once the wolves were no longer considered endangered, permission was obtained to organize the first legal wolf hunt in decades, from September 1, 2009, through the end of the

(continued)

Anti-predator Hunting, *continued*

year. The goal was to "harvest" enough wolves to reduce the population by as much as 30 percent in the affected states. In Idaho where the quota of wolves to be killed was 220, more than fourteen thousand wolf-hunting permits were sold during the first ten days they were available, and the situation was similar in Wyoming and Montana. Wolves are elusive prey, and Jon Rachel, the wildlife manager for the Idaho Department of Fish and Game, commented: "It's clear it's not going to be easy." It remains to be determined whether the legal wolf hunt was the right thing to do, but it seems safe to say this approach will have less environmental impact than poison.

other avian scavengers in West Africa. The U.S. manufacturer of the toxin has agreed to suspend exports of Furadan to Kenya in the wake of the reported poisonings, but it is still widely available and remains a threat to many species of wildlife.

Question 10: Do other environmental toxins endanger birds of prey?

Answer: The Bald Eagle *Haliaeetus leucocephalus* was removed from the federal government's endangered species list in 2006, but a 2008 publication by Maine's BioDiversity Research Institute reported accumulations of mercury in the blood and feathers of Bald Eagle young in the Delaware/Catskill region of New York State that were elevated and caused concern. Wind-borne mercury from coal-burning power plants in the Midwest falls into lakes and streams in the Catskill area and forms methylmercury, which is ingested by worms and other organisms that are then eaten by fish that are in turn consumed by eagles and fed to their nestlings.

The Institute conducted a survey to identify biological mercury hotspots in the northeastern United States and southeastern Canada. A hotspot is a statistically adequate sample from a location that, compared to surrounding areas, has concentrations of mercury in fish, birds, and mammals that exceed human or wildlife health criteria. The survey identified five biological mercury hotspots and nine areas of concern. In some areas, the mercury was due to emissions from local sources that became concentrated in animals. The research noted the need for a monitoring network and the development of mitigation strategies.

In a report published in 2007, the Institute also found Maine's Bald Eagle population diminished. To study mercury concentrations in fish-eating Bald Eagle nestlings and adults in Maine, between 2001 and 2005, researchers collected and analyzed nestling blood, shed adult feathers, and abandoned eggs. The results varied among ten Maine watersheds, and sample sizes were relatively small, but the study did find that blood mercury levels in Maine eaglets were "higher than many regional comparisons, and most similar to populations associated with significant point source pollution problems." (Mercury point-source pollution is usually due to mining or dredging.) Similar results were found in the analysis of the shed adult feathers. Follow ups documented "significant negative relationships" between eagle blood mercury level and the numbers of chicks fledged and occupied nests, suggesting a reproductive impact from the mercury exposure.

Biologist Brian Woodbridge of the Klamath National Forest in Northern California was hired in the early 1990s by the U.S. Forest Service to study Swainson's Hawks *Buteo swainsoni* in Butte Valley National Grassland. Swainson's Hawks can live twenty years or more, and at first the population seemed stable, thriving on the large populations of rodents in the local alfalfa fields and arid grasslands. The Swainson's Hawk population, estimated at 400,000 in North America, nests from about May through August on prairies, plains, and deserts west of the Mississippi River and north into Canada. As he followed the hawks over time, Woodbridge began to notice that in some years fewer hawks returned from migration than would be expected, a decline also reported in several western Canada provinces.

In 1994, using satellite telemetry to follow two hawks from Butte Valley, he discovered that they had joined huge flocks that migrated to winter in the alfalfa and sunflower fields in La Pampa province in Argentina, 300 miles (over 650 kilometers) west of Buenos Aires. The hawks feed on grasshoppers that thrive on the crops. In the winter of 1995–1996 he and his colleagues travelled to Argentina and found more than four thousand dead hawks at four sites in La Pampa. Forensic evidence confirmed that they had died as a result of being exposed to the organophosphate pesticide monocrotophos, used by sunflower farmers to combat grasshoppers. Evidence indicated that some birds died from direct exposure to the pesticide spray, while others died from eating poisoned grasshoppers. Even at that time, monocrotophos was banned in Canada and it was not manufactured in the United States. It had been withdrawn from sale in the United States in 1988.

An international group of governmental and private institutions succeeded in curtailing the sale of the pesticide in the Argentine pampas, enforcing restrictions on its use, and educating local farmers about the available alternatives. From the winter of 1996–1997 until today, only a few incidents of pesticide exposure have been reported in Argentina, and its use is reported to have been reduced to 2 percent of previous levels. In 1999 Novartis announced it would phase out production of monocrotophos, as well as five other highly toxic pesticides.

The responsiveness of citizens, government agencies, and manufacturers is to be applauded. There have been some changes in public policy that increase the chances that the environmental impact of development projects will be evaluated prior to their being considered for approval, but there is a long road ahead before changes take place that guarantee no serious harm will come to wildlife.

SIX

Raptor Husbandry

Question 1: What is meant by "husbandry"?

Answer: "Husbandry" is an old word with several meanings. Originally defined as "care of the household," the term now usually refers to the scientific management and control of natural or agricultural resources. "Raptor husbandry" is the care, medical treatment, and management of captive raptors in facilities for captive breeding or rehabilitation, public display in zoological parks, and falconry. The amount of space an animal needs to be safely housed, breeding strategies, diet and feeding techniques, diagnosis and treatment, and exercise are all aspects of raptor husbandry.

Question 2: How do zoos and rehabilitation facilities house raptors?

Answer: Raptors should never be housed in small cages such as those used to confine parrots, canaries, or finches. A raptor housed this way will destroy itself against the bars of the cage in an attempt to fly off to a higher perch, or to investigate something it sees as potential prey.

Raptors held for display at zoological parks or housed for captive breeding are usually free lofted, meaning they are not tethered. They are housed in very large aviaries with vertical bars on the outside to prevent escape, lined on the inside with soft netting to prevent damage to wing and tail feathers. These aviaries are ideally so large that raptors can fly freely back and forth to the padded perches provided. Shelters, bath pans, and ledges are standard, and the birds are checked daily to insure that they are healthy. Breeding birds are carefully observed to follow the status of eggs and young.

WildCare Incorporated is an Indiana wildlife center that takes in injured and orphaned birds of prey and other animals needing rehabilitation and care. As is typical at proper facilities, they use rectangular raptor enclosures that vary in size according to the size of the bird. An enclosure for a small owl or hawk is 48 feet long, 12 feet wide, and 8 feet high. For a large eagle, the enclosure is more than twice as large at 100 feet long, 50 feet wide, and 20 feet high.

If a raptor has recovered in the care of a rehabilitator after suffering an illness or injury, it cannot simply be released. A weakened raptor will not survive in the wild, and it needs to be conditioned for release by exercising for several weeks in a weathering pen, strengthening its wings and flight muscles. One type of flight pen for this purpose is L-shaped to encourage birds to make left and right turns while in flight, exercising both wings equally. There are special enclosures where appropriate raptors can hunt live mice to hone their hunting skills before being released.

Question 3: What does a rehabilitator do with a sick or injured raptor?

Answer: Birds may be brought to a licensed rehabilitator by Fish and Wildlife officers or by concerned citizens. The rehabilitator is often a board-certified veterinarian or works under the supervision of someone with similar qualifications. The bird to be admitted may have been injured or orphaned, or may appear ill or starving (see chapter 5 for more details about common injuries and illnesses). Starving young birds found by well-meaning people in the fall and winter and brought to a rehabilitation center are usually the surplus population that would normally die off. The resources of a habitat limit its populations, and healthy populations produce excess young (see chapter 1, question 11: How long do raptors live in the wild?).

Raptor illnesses and injuries are often difficult to diagnose. In the wild, showing any sign of weakness often turns the predator into prey, so obvious signs of illness may not appear until a raptor is almost at death's door. The first task of a rehabilitator when birds are admitted is to stop any bleeding and to treat for shock. They are given a physical examination, including eyes and ears, and blood and feces may be sampled and analyzed to check for infections, anemia, and other clues as to why the bird is debilitated. X-rays may be used to detect fractures or the presence of lead or other hazardous material.

Once the problem has been diagnosed, the rehabilitator develops a treatment plan. A bird may need to stay at the facility for only a few days, or it may take weeks or months to restore it to health. Sometimes all it needs are rest and food. Wound management, daily medication, or surgery can be provided if necessary. On one hand, because most raptors are relatively large, some diagnostic and treatment options are available for them that are not always possible with smaller animals. (One cannot even draw blood from a hummingbird, for example, and it is extremely rare to be able to splint its broken bone.) On the other hand, because a raptor must be a supreme athlete to survive in the wild, treatment that would be satisfactory for another animal may not restore a raptor to a level where it will have a reasonable chance of survival if it is released. A raptor with a physical handicap, such as an imperfectly set wing bone, is not considered releasable.

Some facilities report that about half the raptors admitted each year can be returned to the wild, because they have been completely rehabilitated and are able to see well and fly strongly. Because they are bound to be out of practice, raptors are usually released in the spring when there is likely to be an abundance of easily caught young prey. If they have lasting visual or flight problems that make them unreleasable, they might be suitable for transfer to a zoo or other accredited educational facility, or they may be used for breeding. If a bird is in pain and a sufficient level of health cannot be restored, it may be painlessly euthanized according to international standards. Determining the best course of action is one of the tasks of a rehabilitator.

Question 4: What is "imprinting"?

Answer: "Imprinting" is a term for genetically programmed learning that occurs during a specific, sensitive period shortly after birth. It creates a permanent attachment of certain behavior to specific objects that are then treated as parental figures or potential mates. Konrad Lorenz famously demonstrated how newly incubator-hatched geese would imprint themselves on him or on his boots, following him around as they would their mother goose. In birds that leave the nest shortly after hatching, the critical period for imprinting is between thirteen and sixteen hours after hatching. Raptor young remain nestlings for a longer period, and their imprinting occurs over weeks, rather than days or hours. It is part of the same instinctive process by which a human baby learns from its parents how to survive.

A raptor nestling raised by humans will not have had the opportunity to learn how to hunt from its parents, and its prospects for survival in the wild will be dim. If it becomes imprinted on people, it will aggressively seek out humans for food and even as potential mates, and it will not be inclined to mate with its own species. Imprinted birds can be placed in zoos or other educational facilities where their comfort with humans can make them suitable for demonstrating some of the skills raptors possess.

When working with young birds in a rehabilitation facility, care is taken whenever possible not to teach a captive bird to depend on people. A wild breeding pair of the same species may be willing to provide foster care for an orphaned nestling, and if suitable wild foster parents cannot be found, a nurturing male or female captive bird may be able to fill the role. Obviously, the foster parents must be watched very closely to be sure they are not inclined to eat the youngster. Hand-feeding raptor nestlings that are suitable for release requires hiding the hand doing the

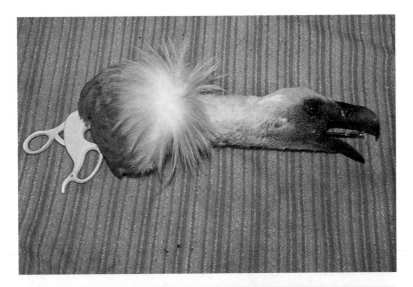

Figure 22. *Top,* a vulture puppet invented by Yigal Miller to feed vulture chicks at the Tisch Family Zoo in Jerusalem, Israel, without their becoming imprinted and dependent on humans for food before they mature enough to be releasable. *Bottom,* feeding a young vulture chick with the puppet. *(Photos courtesy of Michael Erez, modified by Jeff Groth)*

feeding. Techniques include putting a blanket over the handler while he or she introduces food, or darkening the enclosure for a few moments while the food is inserted. A favorite technique involves covering the hand with a sock puppet or other implement that has some resemblance to an adult of the same species as the nestling. Food is delivered in the covered hand through a hatch or hole in the cage wall (see figure 22).

Fledglings can be let go with a soft release, which involves setting the young birds up in a predator-proof nest and continuing to feed them as they come and go and gather strength. This behavior mimics what their parents would do, and they will fly away when they're ready if the timing of the release is right. Jim Parks and Julie Collier, licensed Massachusetts-based rehabilitators, go to great lengths to make sure that when raptors they have cared for are ready for release, they have a good chance to function normally in the wild. Even though these birds have not been imprinted, the rehabilitators "may spend a day or two before a scheduled release banging pots and pans near the bird, or running toward it again and again, yelling madly," so that, for its own safety, it will fear humans.

Question 5: What are raptors fed in captivity?

Answer: A constant source of good fresh meat is necessary to maintain healthy birds in captivity. In warm climates or summer months, refrigeration is an absolute necessity. Some falconers, zoos, and rehabilitators rely on poultry farmers to supply culled, day-old chicks. These can be frozen to provide a steady source of food. Coturnix Quail *Coturnix coturnix* are now commercially available as raptor food, and hundreds at a time can be purchased frozen and shipped for use. Day-old chicks and quail make excellent falcon food, and other types of raptors can be fed on commercially raised mice and rats. Pigeons are sometimes fed to captive raptors, but this practice may carry some health risk for the raptor: some pigeons have been exposed to pesticides or may carry the protozoan *Trichomonas gallinae,* the cause of a serious disease in raptors known as frounce.

Falconers regulate their birds' weight carefully. The birds are weighed every day, and care is taken during hunting season to maintain them at flying weight—lean enough to be motivated to chase quarry, but not so thin as to compromise health. Because of this system, a falconer feeds his birds more or less food each day according to its weight. When falconry birds are "put up" to molt, they are given as much food as they want so that their new feather growth is robust and complete.

Captive raptors on display and recovering from injuries are generally given as much food as they can eat and often get so fat that they can actually skip a day or two of feeding. A varied diet is good for captive raptors, and they should never be given just one type of meat. Captive fish-eating raptors should be allowed to feed on hatchery-produced trout and other appropriate fishes.

Question 6: How are raptors housed for falconry?

Answer: Due to raptors' anatomy, physiology, and temperament, they require special care and handling. Some of the techniques and jargon used in falconry to describe these elements have changed very little in the course of its long history. Falconers house their birds in a mews or hawk house, a building used solely for the keeping of hunting hawks. Falconry birds have a soft leather strap, a jess, attached to each leg just above the foot, joined at the free ends to the top ring of a strong swivel. The bottom ring of the swivel has a longer leather or braided nylon leash attached. The swivel prevents jesses and leash from twisting. The free end of the leash is then tied in a falconer's knot (a knot that can be tied and untied with one hand, since falconers have a bird on the other hand) to a swivel ring on the raptor's perch. Below the perch is a broad shelf, so if the bird jumps off its perch it ends up sitting on this shelf rather than hanging by its jesses. This arrangement is known as a shelf perch. The perch itself is padded with Astroturf, leather, or thick hemp rope, which simulates the uneven, rough surface of a branch or rocky ledge. This surface keeps a captive raptor's feet healthy and in good shape. Several raptors can be kept in one mews, provided they are tethered on perches spaced far enough apart so that they cannot reach each other.

To keep the mews dry and mold free requires good ventilation, a pitched roof to conduct rain away, and dry clay litter or sand on the shelf under the perch. A table with a scale and daily weight chart, as well as some other falconry equipment, can be kept in the mews, but pieces of food should never be left behind because the bird will keep trying to reach them.

The mews protects captive raptors from weather and from nocturnal predators when the falconer is away. During the day hunting hawks are transferred to a weathering yard to get fresh air and sunshine, and they are provided with a shallow pan of water for bathing. Weathering yards are usually large sections of well-kept lawn completely enclosed by fencing or poultry wire. The shape of the bird's perch mimics its preferred perch in the wild: falcons are tethered to block perches, while other types of hawks sit on curved ring or bow perches.

Question 7: How long do raptors live in captivity?

Answer: Zoos and educational programs are the sources of information on how long captive birds of prey live: small falcons and accipiters can live about fifteen years; medium-sized buzzards and kites can live twenty to forty years; and large birds such as vultures and eagles have been known to live for as long as fifty-five or sixty years in captivity, according to statistics gathered by Jean-Marc Thiollay of the Laboratoire d'Ecologie, Ecole Normale Supérieure, Paris. This is more than double their usual maximum lifespan in the wild.

SEVEN

Taming and Training

Question 1: What is "falconry"?

Answer: Originally called "hawking," "falconry" is the general name for the sport of hunting with raptors, one of the oldest sports in the world. The original goal of this type of hunting was to put food on the table. Until the seventeenth century when the popularity of the sport began to wane, hunting with hawks was the favored sport of medieval Europe's aristocrats. Kings traded hawks for prisoners, and knights carried their birds around on the fist, even taking them to church. "True falcons," by which they meant female Peregrines *Falco peregrinus*, were considered so special that no commoner was permitted to own one—named for its most elite bird, the elite sport became known as "falconry." (Falconry birds are still usually referred to as "she," although males are also used.) Commoners hunted with accipiters such as Goshawks *Accipiter gentilis* and Sparrowhawks *A. nisus*, species considered lesser by the aristocracy, even though these birds generally caught game more consistently than falcons did. But accipiters were flown right from the fist, and so did not provide the exciting display of a falcon participating in the hunt from a great height and then diving and colliding in midair with its quarry.

Raptors evoke intense emotions. "To have such a magnificent creature as a hawk, falcon, or eagle accept you as its companion and to allow you to enter its space is to gain a rare perspective

Figure 23. A Gyrfalcon *Falco rusticolus*, soft leather jesses attached to its legs, perches on the gloved fist of its handler as part of a visiting exhibit of raptors in Central Park, New York, from Hawk Creek Wildlife Center. *(Photo courtesy of Carol A. Butler)*

on life," says Stephen Bodio, an avid falconer. According to ornithologist and falconer Tom Cade, the falconer does not do much once the pair arrives at the hunting field; he considers falconry an exciting and ritualistic form of bird-watching. The bird instinctively knows how to kill, and its desire to do so is usually strong; the falconer participates vicariously. The falconer's role is to develop just enough rapport with the wild bird so that she will allow the falconer to approach when she has killed prey. She must be trained to return to the falconer's glove if she fails to make a kill, signaled with a whistle.

The Victorian era's emphasis on respectability and civility caused falconry to fall from favor. With the advent of guns, shooting replaced other methods of hunting for sport and for food. Industrial, residential, and agricultural development in

Figure 24. Hannah Keller and her black Gyr/Saker hybrid. Hood removed, the falcon gets its bearings before Keller casts her off her gloved fist. *(Photo courtesy of Hannah Keller)*

Europe reduced the availability of open land that had been a refuge for game birds and a place to hunt, and by the nineteenth century, falconry was rarely practiced anywhere. The opening of the American West with its huge tracts of land and abundant game led to a new interest in falconry, and today there are avid devotees in many places in the world.

Question 2: Where and when did humans begin using captive raptors for hunting?

Answer: Several cultures where falconry has flourished lay claim to its origins, but artistic, descriptive, or physical evidence of raptors and people together does not confirm that they were hunting partners. Archaeologists have excavated early settlements in the Near and Middle East that date from 10,000 to 8,000 BCE, and they consistently find bones of birds of prey among the human remains, but they may have simply been eaten or kept as companion animals, exotic pets, or symbols of power.

Many falconers believe that hunting with birds began in the Mongolian Steppes of Central Asia when nomadic herders witnessed eagles killing animals. Some herders would have recognized the potential of hunting with eagles, and taming the birds enough to work with them may have been a conscious effort or an inadvertent discovery. The practice of falconry was probably carried from the extensive Mongol Empire to Japan, where historian Ahizato Pito wrote in 1808 that falcons were given to Chinese princes of the Hiu dynasty, which dates from 2200 BCE. There is a fourth-century BCE gold coin from Ancient Greece that shows Alexander the Great (356–323 BCE) with a hawk on his fist, and many later sculptures, paintings, and tapestries from various cultures show this posture, which is associated with hunting.

In Egypt and other Arab countries, the image of the Eagle of Saladin is or has been part of the national emblem. Saladin (1138–1193 CE) was a Kurdish Muslim born in Tikrit, Iraq. At the height of his power, he ruled over Egypt, Syria, Iraq, part of what is now Saudi Arabia, and Yemen. He led the Muslims against the Crusaders, and his heroism and leadership made him a symbol of Arab nationalism. The belief that the eagle was his personal symbol is disputed by some archeologists.

Teaching birds of prey to catch food for the table is mentioned in the Koran, and a treatise in Arabic about falconry dates to the eighth or ninth century. In the arid Arab countries, game is not naturally plentiful, and passage birds are traditionally trapped and trained on their migration south and released in the spring. Captive breeding has flourished in this area because of the nature of the habitat. India has a long tradition of falconry, and there are hawk markets in Pakistan. The sport was eventually introduced in Europe, probably by Germanic adventurers who observed this form of hunting in their travels to the East. The Romans may have learned falconry from the Greeks.

The first known Western book about falconry, *De Arte Venandi cum Avibus* (*The Art of Falconry*, literally, "the art of hunting with birds"), was completed by Holy Roman emperor Frederick II of Hohenstaufen in 1247 CE. An avid hunter and naturalist, he spent more than thirty years gathering material for this treatise.

A

Judging by its engorged crop, this Red-shouldered Hawk *Buteo lineatus* is finished feeding for the day. *(Photo courtesy of Richard Ettlinger)*

A Red-tailed Hawk *Buteo jamaicensis* uses its talons to hold down prey while ripping off bits of flesh with the sharp, curved bill typical of raptors. *(Photo courtesy of Bryn Tassell)*

B

Head and neck ablaze with color, King Vultures *Sarcoramphus papa* with their intimidating size often persuade other vulture species to leave the carcass to them. *(Photo courtesy of Patricia van Casteren)*

With a wingspan of approximately ten feet, the Andean Condor *Vultur gryphus* of South America soars effortlessly on warm air thermals in search of carrion. *(Photo courtesy of Keven Law)*

The Palm Nut Vulture *Gypohierax angolensis* is singular among raptors for consistently eating plant material as well as meat. *(Photo courtesy of Tristan Bantock)*

C

Exceptional among raptors for its hunting style—walking rather than flying—the Secretary Bird *Sagittarius serpentarius* of Africa gets its name from the resemblance of its long head plumes to quill pens. *(Photo courtesy of Ben Lunsford)*

Carrying nest material in its talons, a Swallow-tailed Kite *Elanoides forficatus* displays gracefulness and agility. *(Photo courtesy of Richard Ettlinger)*

This brilliantly patterned Collared Falconet *Microhierax caerulescens* illustrates that not all raptors are large or drab. (*Photo courtesy of Somchai Kanchanasut*)

Unique in the genus *Geranospiza*, the Crane Hawk *G. caerulescens* forages for prey by inserting its specially jointed leg inside tree cavities to snatch nestling birds or sleeping bats. (*Photo courtesy of Ric Fontijn*)

The Crested Caracara *Caracara cheriway* is the only true falcon, family Falconidae, that builds a nest and hunts on the ground. (*Photo courtesy of Richard Ettlinger*)

With short, rounded wings and a long tail, the Ornate Hawk eagle *Spizaetus ornatus* easily maneuvers through the dense tropical canopy in search of prey. *(Photo courtesy of Clayton Bownds)*

The largest eagle in Africa, the Martial Eagle *Polemateus bellicosus* can catch and kill young ostriches and small antelopes with its powerful feet. *(Photo courtesy of Olivier Delaere)*

Named for the adventurer and naturalist George Wilhelm Steller, the Steller's Sea Eagle *Haliaeetus pelagicus* is a denizen of the cold coastal regions of eastern Russia. *(Photo courtesy of Tom Merigan)*

F

A graceful male Snail Kite *Rostrhamus sociabilis* surveys his territory while cruising over a Florida swamp. *(Photo courtesy of Richard Ettlinger)*

This female Snail Kite *Rostrhamus sociabilis* is removing an apple snail *Pomacea paludosa* from its shell after severing the tough attachment muscle with its highly modified upper mandible. *(Photo courtesy of Richard Ettlinger)*

G

Silent and ghostlike, a Snowy Owl *Bubo (=Nyctea) scandiacus* flies low over the frozen tundra in search of lemmings to feed its growing chicks. *(Photo courtesy of David Hemmings)*

Social and comical looking, these Burrowing Owls *Athene cunicularia* are at home in the abandoned burrows of ground squirrels. *(Photo courtesy of Eduardo López)*

The secretive and elusive Spot-bellied Eagle-owl *Bubo nipalensis* is known as the Devil Bird in Sri Lanka due to its bizarre, humanlike call. *(Photo courtesy of Somchai Kanchanasut)*

H

Regardless of the best intentions of humans, technology often proves lethal to raptors. Here, the blades of a wind turbine have killed a Red-tailed Hawk *Buteo jamaicensis. (Photo courtesy of Charlene Burge)*

Figure 25. Many societies have incorporated raptors into their art and culture, for example, this apparently stylized vulture carved by an Ashanti in Ghana, West Africa. *(Photo courtesy of the Joel Cooner Gallery, Dallas, Texas)*

It contains scholarly, instructive, detailed descriptions of his experiences with birds of prey, as well as lavish illustrations of animals, people, and falconry equipment. In the same period, Marco Polo reported that the head of the Mongol Empire, Chinese emperor Kublai Khan (1215–1294 CE), grandson of Genghis Kahn, went on hunting expeditions "with 10,000 falconers and some 500 Gyrfalcons, besides Peregrines, Sakers, and other hawks in great numbers." The hunters reportedly took along a pavilion lined in beaten gold and borne by four elephants.

From about 500 to 1600 CE, European aristocrats and kings practiced falconry extravagantly and competitively. A well-trained falcon was a suitable present for a lady or gentleman, and its presentation and first flight were grand occasions. A story often repeated tells of twelfth-century England's Henry II and his nobles taking their hooded falcons with them to dinner. When specially prepared meat pies containing small live birds were opened during the meal, the falcons' hoods were removed and they supposedly set upon the birds.

Women in Falconry

Most people probably think of falconry as a man's sport, but women have been enthusiastic practitioners over the course of its history and are the fastest-growing demographic in falconry today, according to Frank Bond, president of the International Association for Falconry and Conservation of Birds of Prey (IAF). James Strutt in 1801 noted that English "ladies not only accompanied the gentlemen in pursuit of the diversion [falconry], but often practiced it by themselves; and even excelled the men in knowledge and exercise of the art." Historical treatises often mention well-known women falconers.

It is reported that the controversial Mary, Queen of Scots (1542–1587), described as "handsome, brave, and proud," was allowed to fly her Merlin *Falco columbaris* from her window when she was in prison awaiting execution. Although Queen Elizabeth I had ordered her execution, perhaps she was sympathetic because the queen herself was a falconer. Some sources say she appointed a woman, Mary of Canterbury, to the position of Grand Master of Falconry. Empress Catherine of Russia (Catherine the Great), who reigned from 1762 to 1796, also flew a Merlin, historically the bird most popular with women.

Dame Juliana Berners, prioress of the Sopwell nunnery near St. Albans, England, was interested in falconry, although there is no mention of her engaging in the sport; she is credited with writing the parts of the 1486 *Boke of St. Albans* that had to do with "haukynge," "huntyng," and "fysshyng." (Her treatise on fishing is available on the Internet.)

According to Noriko Otsuka of the Japanese Falconers Association, the *Nihon Shoki* (Annals of Japan) reports that in 355 CE, Sakenokimi, an immigrant from what is now Korea, trained a hawk given to Emperor Nintoku. The emperor was so enthusiastic about this new way of hunting that he established the Hawkers' Guild, which began a long, aristocratic tradition in Japan of *takagari* (falconry). According to Otsuka, in order to retain another immigrant falconer, Kanemitsu,

Women in Falconry

Emperor Nintoku gave him Kochiku, a beautiful woman from the court. Other sources say that Kochiku was the first Japanese falconer, and that her daughter, Shukou, was also a practitioner of the sport. Japanese documents from the ninth century are frequently cited as reporting the presence of women falconers.

Today, approximately 10 or 11 percent of the active falconers in the world are women, according to an IAF estimate and the North American Falconers' Association. Women historically have flown Merlins, but IAF president Bond has observed that increasing numbers of women are flying larger species of raptors. Elizabeth Schoultz, for example, is a New York/Pennsylvania falconer from Michigan who aspires to fly an eagle; she identifies herself as a huntress.

Master falconer Anne Price has observed that men who get interested in falconry today have generally been hunting all their lives and come across falconry as another way to hunt, while women tend to get involved with raptors through caretaking as educators or rehabilitators, and their interest in falconry develops as they learn about the birds. Her view seems to be borne out by falconers like New Zealander Debbie Stewart, who first became involved with raptors when she worked as curator at Rainbow Springs, a tourism facility where people brought in injured birds. She is now involved in research and the captive management of native falcons in New Zealand, and is a founder of the Wingspan Birds of Prey Trust in Rotorua, New Zealand. Seattle falconer Lydia Ash is a licensed rehabilitator, and Hannah Keller is a licensed falconer who works in a raptor-breeding program in Las Vegas, Nevada.

It is not uncommon for a woman to develop an interest in falconry because a family member or male partner has an interest in the sport, not unlike the old days. Jemima Parry-Jones came from a falconry family, taking over from her father and directing raptor centers in Britain and in the United

(continued)

States. Emma Ford, along with her husband, Steve, was invited by the president of the United Arab Emirates to study the classic Arabic method of hawking in Abu Dhabi. In 1982, the couple founded the British School of Falconry, and Emma Ford has subsequently acquired an international reputation as a speaker, teacher, author, and governmental advisor on subjects related to falconry. Virginia falconer Jill Morrow reports that she got into falconry when she met her husband.

Some contemporary women who are falconers report getting interested in falconry as young girls. Jemima Parry-Jones had her first bird, a kestrel, when she was six years old. The author Helen Macdonald, who has worked in falcon conservation in Britain and the United Arab Emirates, had a kestrel that was permitted to roost in her bedroom when she was a "small girl." Anne Price, master falconer and curator of raptors at the Raptor Education Foundation in Colorado, discovered her passion for raptors at the age of twelve and received her falconry license at age sixteen, and Elizabeth Schoultz became fascinated with a Harris's Hawk *Parabuteo unicinctus* brought into her seventh-grade classroom on the glove of a classmate's father. Eventually she found her way to Emma Ford's falconry school and obtained her falconry license.

Karen "Kitty" Tolson Carroll, a licensed falconer since 1974, lectures, writes, consults, and gives presentations with hawks and falcons. She says her interest in falconry was stimulated by the movie *My Side of the Mountain,* the story of a boy and a falcon surviving together on a mountain that is based on the book by Jean Craighead George. George was herself born into a family of naturalists, and her own first pet was a Turkey Vulture *Cathartes oura.* Wildlife researcher and falconer Fran Hammerstrom, author of *An Eagle to the Sky* and other books, was an ornithologist and a raptor bander. Master falconer, breeder, and bander Bernadette Richter runs the rehabilitation facility Save Our American Raptors (SOAR) in Illinois with her husband. Veterinarian Meg Robinson of the Waterford Veterinary Clinic in Wisconsin is a lifetime falconer specializing in goshawks.

There is an interesting feminine term used in falconry, "madonnaing" coined by Fran Hammerstrom. When an eagle or any large raptor has been trapped in the wild, it is customary to swaddle the body and feet and to hood the head to immobilize the bird—held lying on its back, it is madonnaed (see figure 26). Since the trapping often occurs a long way from where vehicles are parked, it is a practical way to carry the bird so it does not struggle. It might look like the bird is being comforted and even cuddled like an infant, but more likely the bird is simply surrendering to the forced immobility.

The number of women practicing falconry in North America is probably greater than the relative number of women who hunt in other ways, perhaps because, as falconer Anne Price speculates, "falconry has an inherent husbandry aspect to it that requires year round care and patience, which women typically possess in greater quantities than men. When duck season is over, you can clean your gun and put it in the cabinet, but . . . that bird still needs daily attention." Today, any woman can study, apprentice herself to a falconer, and take a licensing exam that will be her passport to participation in the sport.

Figure 26. Falconer Elizabeth Schoultz demonstrates how to safely transport a newly trapped eagle. She has hooded the bird and gently wrapped its legs and body with gauze and soft cloth to prevent it from struggling and injuring itself or her. (*Photo courtesy of Elizabeth Schoultz*)

In the fourteenth century, perches were commonly installed in the rooms of inns so that hawks would have a proper place to rest while their owners went to dinner. The glove on which a falcon perched and the leather jesses around its legs were often decorated and sometimes embroidered in gold. Jewels and elaborate arrangements of feathers decorated the leather hoods used to cover the birds' eyes, and they wore bells made in specialty shops in Pakistan and Milan. The bells were commonly embossed with the owner's crest and rang with a distinctive sound so the owners could identify their bird.

The son of Phillip the Bold was captured in 1396 at the battle of Nicopolis against the Ottoman Empire in the last large-scale crusade of the Middle Ages. Phillip, Duke of Burgundy from 1363 to 1404, was the younger son of King John II of France. He is said to have sent twelve rare white Gyrfalcons *F. rusticolus* and a jeweled gauntlet (paid for by Carl VI of France) as gifts to Sultan Bajazet (also spelled Beyazid and Bayezid), who held his son. The Ottoman sultan reportedly maintained seven thousand falconers in his hunting establishment. Although he had previously refused a large ransom payment for the duke's son, the sultan was reportedly so pleased with the beautiful white falcons that he ordered the son released.

In the United States, remnants of falconry's extravagant history appear in the decorations on some gloves and hoods. The wreck of a Spanish ship dating back to the conquistadores was raised in 2009 in U.S. coastal waters, and it contained pairs of falconry bells made in Pakistan, suggesting that the conquistadores may have brought trained raptors to the New World. There are stories claiming that Hernán Cortéz saw trained birds of prey used for falconry in the court of King Montezuma, and there is evidence that Native Americans had a traditional relationship with eagles and other raptors, although there is no clear evidence of falconry. Falconry never became wildly popular in the New World because the middle class held in great disdain anything practiced by the European aristocracy.

By 1900, records of the practice of falconry started to appear in the United States, and the number of practicing falconers slowly increased. An article on falconry published in the December 1920 issue of *National Geographic* magazine, written and illustrated by the celebrated bird artist Louis Agassiz Fuertes, has been credited with stirring Americans' interest in the sport. In 1934, the first falconry club was formed—the Peregrine Club at the University of Pennsylvania—and the North American Falconers' Association was established in 1961. At present, there are estimated to be 4,500 falconers in North America, about 10 percent of whom are women.

Question 3: What role do dogs play in falconry?

Answer: Frederick II of Hohenstaufen (1194–1250), Holy Roman emperor from 1120 until his death, was an avid falconer. In

Figure 27. Hawks and hounds wait patiently for falconers to begin the hunt, when the dogs will find and point game while the falcon circles high above. At the right moment, a signal from the falconer sends the dog in to flush the game and the falcon stoops in for the kill. *(Photo courtesy of Tim Gallagher)*

his famously practical book on falconry and hawking, he recommended that a dog used in falconry should be from a lineage bred for this purpose for generations. The primary roles of the dog are to point when it spots game and to flush game birds on command so they become targets for falcons to attack in flight. His ideal dog would have thick hair to resist cold and dampness and be big enough to see over grassy cover, but not so big as to endanger the falcon. It would be agile, not afraid of water, and, of course, trainable and obedient. He provides instructions on how to train the dog and falcon to trust each other and to cooperate, explaining that the falcon should learn to defer to the dog.

Many modern hunters agree that most birds of prey are capable of quickly learning to work with a well-trained dog once they realize that the dog increases their chances of making a kill. A coordinated dog and bird team working together reduces the human member of the team to the role of spectator and cheering section, a common observation made by falconers about this type of hunting.

Question 4: How and where can you acquire a raptor to train?

Answer: Only individuals holding the proper permits can legally acquire and possess raptors to train for falconry (see chapter 9, question 5: Are any raptors endangered?). Some raptors are taken as nestlings, called "eyases," and some are passage birds, young birds trapped on their first autumnal migration using techniques that have changed little over the centuries. Knowledge of the birds' breeding behavior and typical nest site selection is necessary to take nestlings for training. Passage birds are usually trapped along barrier beaches or mountain ridges that are natural corridors or flyways. Birds use warm air thermals to soar and conserve energy, and prey species are usually abundant and often are migrating at the same time.

A relatively new way of acquiring raptors to train is to purchase captive-bred birds. Since about 1980, captive breeding

My First Hawk

I was lucky that when I was a young child my family moved from New York City to the tiny village of Patchogue on rural Long Island, when it was still bordered by mixed deciduous woods to the north, west, and east. The Great South Bay laps at the southern border of the village, and a few old salts still raked clams from the bottom of the bay. As we grew up, my friends and I ran trap lines in the salt marshes and caught muskrat *Ondatra zibethicus,* raccoon *Procyon lotor,* and an occasional red fox *Vulpes vulpes.* We fished the bay for eels and flounder, and a few of us angled for trout in the freshwater streams near my house. By the time I was twelve, I was hunting Bobwhite Quail *Colinus virginianus,* Ring-necked Pheasant *Phasianus colchichus,* and cottontail rabbit *Sylvilagus floridanus.*

I had a pigeon loft in my yard, and one fall day a hawk exploded from the surrounding woods and killed one of my fantails. Transfixed, I watched the hawk pluck and eat the pigeon on the lawn. Then and there I decided to learn about raptors, and I started to read everything I could get my hands on at the local library. I borrowed a copy of Humphrey Ap Evans's *Falconry for You* and Phillip Glasier's *As the Falcon Her Bells.* These classics fascinated me with the idea that raptors could be trained to hunt with people, but they provided little information with regard to how one could acquire a hawk in the United States, something I was becoming determined to do.

The cover of *Field and Stream* always caught my eye whenever my father or older brother took me to a store that displayed it on the magazine rack. While waiting for them one day, I idly flipped through the pages and saw an advertisement declaring that there were hawks for sale for falconry from a dealer in Florida. I wrote a letter to the dealer as soon as I got home, and when the price list and catalog of raptors arrived, I spent weeks gazing at the tiny, glossy, black and white photos of hawks, falcons, eagles, and owls, each with a description and

(continued)

a price. I vividly recall the photo of a small boy holding what seemed to be a huge Red-tailed Hawk *Buteo jamaicensis* on his gloved hand, and I decided that this was the type of bird I would purchase for my own. Now I had to convince my parents and siblings that I should have one, and I had no idea what I was going to feed this monster or where I would house it. Everyone agreed that if I could come up with the sixty dollars for the bird, I could have it. My paper route was netting me about twenty dollars a week, so after three weeks I slipped sixty dollars cash into an envelope with the order form filled out for one Red-tailed Hawk.

Although it seemed like years, I waited only about a week before receiving a phone call from the freight department at Islip McArthur Airport near my home. We drove over to pick up the bird and were asked if we possessed wildlife permits. We said no, but somehow we were given the box anyway. It was huge, about three feet high, and we all nervously imagined that the predatory creature inside completely filled the space. I had prepared a large holding pen for the hawk, and when we got home, I carefully opened the top of the box so I could transfer the bird to its new home.

I wore thick leather welding gloves borrowed from my brother's friend who insisted that we had to protect ourselves from "the killer hawk." As light slowly illuminated the interior of the box, it became clear that the gloves were as unnecessary as our trepidation, because hunched at the bottom was a bantam-sized, ragged brown bird with a hooked beak. I gently lifted the sorry creature out and placed it in the pen. Its body was all dark brown with rust-colored shoulder patches and tarsal feathers. Its almost black tail sported a white terminal band and rump patch, and its face was curiously bare and bright yellow. Glancing at the invoice, I discovered a note: "We apologize. We are out of Red-tailed Hawks. We have sent you a Red-shouldered Hawk *B. lineatus*." I located some paintings of Red-shouldered Hawks but they looked nothing like my bird,

My First Hawk

and while we settled in with the bird, I kept wondering what kind of bird they had sent me.

Judging by the bird's size, I surmised that it was a male. We jessed it up (affixed a leather jess or strap to each leg), according to the instructions in my falconry books, and placed it in the basement on a specially designed perch. I fed my bird beef chuck and chicken, and my pigeons gradually disappeared into my new raptor's crop. For the next several weeks I manned the bird, getting it accustomed to being around people. I trained it to jump to my fist for bits of meat, and we had limited success getting it to chase squirrels and rabbits. One day I accidentally stumbled upon Audubon's painting of an adult Harris's Hawk, plate number 392 in his *Birds of America*. There on the page, standing alone on a bare branch, was my hawk, a *Parabuteo unicinctus*, rendered by Audubon in the 1800s. I was elated to finally clinch the identification of my hawk and get a brief description of its habits.

I kept the bird for two years, until one day I came home from school to find him extremely ill. He could not stand, and his eyes were half closed. I tried to nurse him and gave him food and water, which he refused; shortly afterward he died. Working with this Harris's Hawk so long ago and learning about its life and its use in falconry has influenced the entire course of my life. Raptors, in particular, are an inextricable component of my existence. I have crossed paths with many raptors since my formative years, and in a very real sense I owe that pleasure to that first Harris's Hawk on Long Island.

—Peter Capainolo

and sale of raptors by falconers has become common, and many falconry equipment catalogs and falconry trade journals routinely run ads from breeders who produce high-quality, healthy peregrines, Merlins *F. columbarius*, Gyrfalcons, and all sorts of hybrid falcons, as well as Goshawks and other accipiters.

Captive-breeding permits are required to do this, and falconers still must be licensed in order to purchase a bird.

Because the legal trapping and training of a wild raptor is an experience like no other, it seems likely that many falconers will continue to acquire their birds this way, in addition to occasionally purchasing captive-bred birds.

Question 5: Do you always need a license to possess a raptor?

Answer: Raptor rehabilitators and museums must have special permits to receive and possess raptors, whether they are alive, dead, or injured. Private citizens who find an injured raptor cannot legally keep the bird and must give it to a licensed rehabilitator or veterinarian. If it has expired, it can be given to a museum. Purchasing a raptor that is protected under the Migratory Bird Treaty Act requires both a federal and state permit to train and use the bird for falconry or education. If you purchase an "exotic" raptor, which means one *not* listed under the act, you do not need a federal permit but you might need one issued by your state.

If you are going to use any bird to hunt, you need a federal falconry permit. Hybrid falcons need to be on a permit if they have been bred from a protected species, and captive-bred raptors are strictly regulated and have become a significant source of raptors used in falconry. Limited numbers of non-endangered raptor species may be acquired from the wild for falconry, but their possession is regulated by the federal and state governments. Some states require that falconry birds be fitted with a special band, not to be confused with bands used for research.

Falconry is legal in all states except Hawaii. In many states, a falconer must hold both a federal and a state falconry permit, as well as a hunting license appropriate for the game being hunted that specifies the local hunting seasons and bag limits. In New York, for example, there are three classes of falconry permits: apprentice, general, and master. An apprentice must be at least fourteen years old, pass a written exam, be sponsored by a general or master falconer, and pass a facility and equipment inspection by a conservation officer. An apprentice may become a general falconer after practicing under a sponsor for two years, and may become a master after practicing at the general level for five years. Courses are offered to expose people interested in falconry to the practical aspects of owning a bird of prey, and to give them some idea of the responsibilities involved in owning and training a bird of prey.

Keeping in mind that laws and requirements may change at any time, it is interesting to consider some approaches to falconry regulation in other countries for purposes of comparison. In the Netherlands, the Flora and Fauna Act, which became effective in 2002, includes falconry and is quite restrictive. Only the Goshawk and the Peregrine Falcon can be flown, and at only five prey species: brown hare *Lepus europaeus,* European rabbit *Oryctolagus cuniculus,* Mallard *Anas platyrhynchos,* Wood Pigeon *Columba palumbus,* and Ring-necked Pheasant *Phasianus colchicus.* To qualify for one of the two hundred hawking licenses issued annually, one must pass an examination that has both theoretical and practical components. There are some grandfathering provisions for experienced individuals who have a proven record, a previous license, and experience with a mentor from a recognized hunting club.

To practice falconry in Germany one needs a license, a *Falknerjagdschein,* and must pass hunting and falconry examinations that include knowledge of raptors, laws protecting them, raptor husbandry, and the practical aspects of falconry. Only three species of the fifteen native German raptors are permitted in falconry, and they must be banded and aviary bred. Goshawks are the most commonly flown hawks (60 percent), followed by Peregrine Falcons and Golden Eagles *Aquila chrysaetos.* An owner may have no more than two birds, and permission must be obtained to hawk in a hunting district, something that is said to be difficult and expensive to accomplish.

Falconry is not illegal in Malaysia, but it is essentially non-existent and is indirectly restricted. No license is required to become a falconer, but the 1972 Protection of Wildlife Act prohibits unlicensed game hunting, and ownership of raptors requires special permits that are said to be nearly impossible to obtain. In some countries, such as Australia, falconry is not legal.

Question 6: Can all raptors be tamed and trained?

Answer: Like dogs and horses, birds of prey vary in their receptiveness to training. For centuries, humans have been taming and training raptors, mostly falcons and hawks. If they are not captive bred and purchased from a licensed dealer, most trained birds are trapped as passage birds. These young birds on their first migratory flight are likely to be easily trained and, using some simple techniques, can be reverted back to a wild and independent condition should a falconer wish to release them. Some species are taken as nestlings because their high-strung temperament makes it very difficult to train them once they reach adulthood.

Extremely temperamental species, such as the African Black Sparrowhawk *A. melanoleucus* and the Cooper's Hawk *A. cooperii*, are infrequently tamed because an unusual degree of tenacity and patience is required for success. Passage accipiters generally do not tame down quickly. They may fly (bate) off the fist over and over again while the falconer attempts to get the bird accustomed to its new surroundings. Often they refuse to eat off the fist for several days, prompting worries about adequate nutrition, only to suddenly begin eating as if they had been tame all along. So independent are these raptors that even a well-tamed and well-trained individual might decide to fly off forever during a hunt. Conversely, buteos and falcons quickly become comfortable on the fist and accept food, and they are much less likely to head for the hills without looking back as an accipiter might.

Only very experienced falconers attempt to handle Golden Eagles, extremely large and powerful birds (see sidebar "Rehabilitating a Golden Eagle"). There are regulations that apply to

Figure 28. A hooded Peregrine Falcon *Falco peregrinus* sits quietly on a cadge, a portable padded perch, before a lecture on birds of prey. Other pieces of the falcon's furniture include a soft leather jess around each leg, a swivel attaching both jesses to the cadge, and a bell around its right leg. The bell helps locate a bird that has chased quarry or wandered out of sight of the falconer. *(Photo courtesy of Carol A. Butler)*

taming and training them, and certain species are totally off limits (see this chapter, question 5: Do you always need a license to possess a raptor?).

Some raptors are not trained because of their biology. Vultures, as an example, may be tamed for exhibitions or wildlife shows, but they are not trained for hunting because they typically feed on carrion and rarely hunt live prey. Generally, only diurnal birds are trained. Since most owls are nocturnal, they are not usually trained because they are not temperamentally suited to daytime flight. The Osprey *Pandion haliaetus* dives for fish and flies to a tree limb to eat its prey, a behavior pattern that would not adapt to training. The Snail Kite *Rostrhamus sociabilis* eats only snails, which would make for a very odd hunting experience.

Question 7: How are raptors trained?

Answer: Training a raptor is an intensive process. The natural instinct of a raptor is to capture prey, and it is the falconer's job to slightly modify the raptors' wild behavior and work with its hunting instincts. During the training process, falconers use a signal, such as a whistle, whenever they give the bird a piece of food, and she is eventually coaxed to fly to the falconer's glove when the signal is given. Most raptors learn this behavior quickly and soon return to the falconer from long distances.

Once the bird has been acquired, it needs to be "manned," a term that describes the process (regardless of the gender of the falconer) used to acclimate a trainee raptor to its new surroundings. The bird has a short leather strap (a jess) attached to each leg, and it usually sits on the gloved left fist of the falconer, who holds the ends of the jesses to keep the bird from flying away. The bird is gradually and gently exposed to small groups of people, hunting dogs, vehicles, and other stimulation. A tidbit of meat is always available on the glove, and eventually even the wildest hawk begins to realize that the falconer means no harm and is a source of food. When the literature describes or pictures medieval aristocrats going about with a bird on the glove—for example, taking it to church—it is because they are manning the bird. A modern falconer watches television or reads with his or her falcon on the glove, spending as much time as possible with the bird. The bird is stroked, at first with a feather or a pencil, and as it becomes more relaxed on the glove, with the ungloved hand. Falcons and buteos are naturally tamer than accipiters, and the high-strung birds require much more time and effort to be tamed down.

Weight control is central to learning when the bird is motivated to hunt. The bird is weighed as soon as it is received, and the weight carefully recorded. The bird is fed only when on the glove, and at first it will not take any food. The weight at which it shows interest in feeding is important, because it is a sign that the level of hunger at that weight can motivate the bird to hunt. It also is the beginning of learning that the falconer is its only source of food. Once the bird takes food from the glove, a bit of food is put into the hood and offered to the bird. When she learns to take that morsel, the hood is slipped onto the bird's head and quickly removed. The food is used in this way to accustom the bird to the hood so that it will remain calm when being transported, possibly thinking it is nighttime. Coauthor Peter Capainolo has ridden on the Long Island Railroad many times from Long Island into Manhattan with a hooded falcon sitting calmly on his fist. The hood also keeps the bird from attacking other birds, if several are on perches in the same location. The term "hoodwinked" means to be deceived or blinded, and we think its origin may be from the practice of hooding falcons.

The next step is to have the bird fly to the fist from a perch outdoors to which it is attached by a long braided cord (a créance) connected to the ends of the jesses. They are connected with a swivel to allow maximum freedom of movement on the tether. The falconer increases the distance from the bird, always with a morsel of meat as a reward on the glove. Eventually it is time for the first free flight, unattached to the créance, with the falconer holding his or her breath in fear that the bird will fly away. The bird by this time has a bell and possibly a lightweight radio telemetry device attached to a leg or to the base of the central tail feather. If it does not come to the glove, it most likely can be located and retrieved.

In the final stage of training, an artificial lure resembling a rabbit or game bird is tossed out to the raptor to simulate the flushing of game in the presence of the falconer. It is then that the avian trainee rapidly realizes that its human companion is an asset in finding quarry. When the bird has reached this point of readiness, it is prepared to accompany the falconer on a hunt. The falconer, perhaps with the assistance of hunting dogs, attempts to flush game for the bird. A trained hawk follows along from tree to tree or branch to branch, while a falcon behaves differently, "waiting on," soaring high above the falconer. If game is found and the raptor makes a kill, a properly trained falcon sits on the quarry while the falconer approaches and calmly steps onto the glove for a reward of meat. In essence, the trainer

Figure 29. Scott Timmons lure flies a Peregrine Falcon *Falco peregrinus* to keep the bird fit. A falcon can also be called down to the lure if it has missed its quarry or has decided to sit on a nearby perch. The falconer blows a whistle each time the falcon is allowed to snatch the lure, which has a small morsel of meat tied to it. *(Photo courtesy of Nick Dunlop)*

works with the bird to facilitate what occurs in nature—the raptor finding and killing prey. The difference is that the raptor learns to be responsive to the presence of its trainer.

These instructions on training a hawk, which hold true today, come from an 1874 Persian treatise on falconry by Husam al-Dawlah Timur Mirza entitled *The Baz-nama-yi Nasiri:* "The more familiar you make your hawk, and the keener you make her on the lure, the better. Now, if you have trained your hawk in less than forty days, you have hurried her training, and hurry is of the Devil, but Deliberation is from God. Be not overhasty or you will spoil her. Such and such a falconer is sure to vaunt his skill, boasting that he has trained and flown his hawk in 15 days. He has erred and blundered: he is not a lover of a hawk but a lover of the pot."

Coauthor Peter Capainolo worked with other falconers for Falcon Environmental Services (FES) at Kennedy Airport, where they regularly flew falcons to chase local birds away from the runways to prevent the birds from colliding with the airplanes (see sidebar "Bird Strikes"). He remembers a "sakret," Mr. Bean, a little male Saker Falcon *F. cherrug*, that was so impatient he would fly out from the fist, turn immediately, then come in and whack into you before you could even get the lure out of your hawking bag. Some people avoided flying Mr. Bean when they could, but he was not as feared by some of the better, quicker lure flyers in the crew. Most raptors trained for falconry soon become acclimated to their handlers and behave more predictably, with a mutual trust and bonding established as the hunting relationship becomes routine.

Question 8: Can a trained raptor be released back into the wild?

Answer: Whether or not a trained raptor used for falconry is releasable depends on how the bird was acquired. If an eyas has been taken from the nest and has been cared for and fed by humans, it is likely to have become imprinted and therefore unreleasable (see chapter 6, question 4: What is imprinting?). Birds raised this way will never fear people, will remain rather aggressive, and generally do not hunt as well as birds taught by their natural parents. Some species of raptors are very difficult to tame and train unless they have been imprinted, but the disadvantage of imprinting is the impossibility of releasing them back to the wild.

Raptors raised by their natural parents and captured on their first migration are known as passage hawks. They have naturally acquired good flying and hunting skills and are leery of people, but they can be tamed and trained and make excellent birds for falconry. Because they retain their wild nature, they can easily be released back into the wild by keeping their weight up and letting them gradually become independent of the falconer (conditioning a captive raptor for release to the wild is known as "hacking"). Even if accidentally lost, passage birds will likely survive because of their hunting experience and the care they received while in captivity. Some falconers intentionally release passage birds after hunting with them for several years in order to return them to the breeding population and to make room for a new hawk to train.

Question 9: What is the status of falconry today?

Answer: Interest in falconry is growing in countries around the world, and many people feel that *this* is the golden age of falconry. The United States has approximately 4,500 active falconers, and we are considered by many of the world's falconers as the "Mecca" of traditional falconry where it is practiced in its purest form. The International Association for Falconry and Conservation of Birds of Prey (IAF) currently represents seventy falconry associations from forty-eight countries, with a total of about thirty-five thousand members worldwide, including the North American Falconers' Association, the largest IAF member organization. China has a long tradition of falconry, particularly among eagle falconers in Central Asia, but they have only recently begun to become accessible to international falconry. If Chinese falconers are included in the worldwide total, it probably exceeds fifty thousand, according to IAF president Frank Bond.

The IAF recently was one of the first nongovernmental organizations recognized by the UN Educational, Scientific and Cultural Organization (UNESCO) as an advisory body for the recognition of "intangible cultural heritages," specifically falconry. The United Arab Emirates are leading the effort to have falconry recognized as an intangible cultural heritage, which would protect it for the foreseeable future. Falconry qualifies for this status in many countries because it meets the standard of having a tradition going back at least two or three generations. The United States would qualify in that regard, but since it is not a signatory to UNESCO, it cannot submit an individual request. An action plan required in order to receive recognition must include legal protection, as exists in the United States on the federal level and in forty-nine states. Another qualification is that falconry's importance must be recognized by an organization like the Archives of Falconry in Boise, Idaho, the most significant repository of falconry art, books, memorabilia, and artifacts in the world. There must also be a firmly established conservation program led by falconers. According to IAF president Frank Bond, a joint UNESCO submission by several nations on behalf of falconry will be made in the near future. Petitioners will include the United Arab Emirates, along with France, Belgium, the Czech Republic, Slovakia, and Hungary.

The sport of falconry has its detractors. Some people are philosophically against hunting of all kinds and see falconry as just another blood sport to be shunned. Others are opposed to keeping such regal creatures as raptors in captivity, even though all domestic animals are descendents of wild and beautiful progenitors. People who are not closely in touch with the natural world may believe that all raptors are on the brink of extinction and that taking birds for falconry poses an unacceptable risk. When educated about these issues and shown data indicating that falconry poses no threat to populations of wild raptors, this latter group often comes to realize that falconry is one way of having a relationship with nature that is active rather than passive.

Because of the extent of human development, most raptors must share their habitat with humans, or their habitat has been changed by farming, forestry, construction, or pollution. To be a successful falconer requires one to be knowledgeable about wildlife ecology and management. Early pioneers in Peregrine breeding and restoration programs like the Peregrine Fund and the Predatory Bird Group in Santa Cruz, California, provided a wealth of experience that has expanded internationally, leading to the development of programs to rescue other species. Individuals interested in falconry such as Heinz Meng, professor emeritus at the State University of New York College at New Paltz; James Enderson of the Colorado Division of Wildlife; Richard Fyfe of the Canadian Wildlife Service; and Tom Cade, professor emeritus at Cornell University, have created strategies to preserve wildlife resources for the future that will help raptors withstand continued development and the possible effects of climate change. Falconry is a natural method of predation and an important cultural heritage, and we hope it will one day be so designated so its traditions may be permanently preserved.

EIGHT

Raptors and People

Question 1: Have attitudes about raptors changed over time?

Answer: Until the eighteenth century, birds were objects of interest mainly because they provided a good meal. In 1758, botanist Carl Linnaeus published *Systema Naturae*, in which he cataloged more than twelve thousand plants and animals. He included seventy-five birds from North America that had been described and named by Mark Catesby, an Englishman, who had gone to the British colonies in North America in 1712. Catesby had attempted to compile an extensive natural history of the American Southeast by collecting, painting, and shipping live specimens home to England.

In 1789, the Reverend Gilbert White in southern England published his own detailed notes as *The Natural History of Birdwatching*. These efforts, along with work by other contemporaries, stimulated an interest in birdwatching as a gentlemanly form of study, which in the nineteenth century developed into a mania to collect and display all sorts of specimens (see chapter 4, question 10: What do raptor eggs look like?). There were still some detractors who condemned bird-eating raptors as "really bad cannibals," the phrase Mabel Osgood Wright chose in her 1847 book *Citizen Bird*.

Frank M. Chapman was a banker who became caught up in the fascination with the natural world. Birding became a serious interest for him and in 1888 he joined the staff of the American

Museum of Natural History in New York, eventually becoming curator of birds. In 1886, on two afternoon strolls down Fourteenth Street in Manhattan (at that time the main shopping avenue), he recorded 700 women wearing hats, and 542 of the hats decorated either with feathers or with entire birds, as reported by Scott Weidensaul in *Birds of a Feather.* Chapman identified forty species of birds on the hats, ranging from songbirds to a Saw-whet Owl *Aegolius acadicus.*

By the early part of the twentieth century, ideas about conservation and the value of preserving species had begun to capture the public interest. Nature study flourished, but raptors remained unredeemed in the public eye in some quarters. Scott Weidensaul quotes ornithologist George Miksch Sutton in 1928: "The Goshawk is a savage destroyer of small game and poultry. His smaller cousins, the Sharp-shinned Hawk and Cooper's Hawk are killers. The Great Horned Owl is destructive at times."

But the tide against raptors was turning, and people began to advocate for the protection of birds of prey, few more energetically than Rosalie Barrow Edge. She had started bird-watching in New York's Central Park in the early 1900s and became an ardent conservationist, eventually adopting as one of her projects an area in the Kittatinny Mountains of Pennsylvania. The locale she chose was extremely popular with hunters in the early part of the century—ideal for shooting birds as they migrated past on the updrafts generated by the configuration of the ridge.

In 1932, Richard Pough, who later founded the Nature Conservancy, photographed rows and rows of dead falcons, hawks, and eagles that he found just below the favored shooting area in those mountains. He used the photos and his writing to gather support for making the ridge off limits to hunters. In 1934, conservationists led by Rosalie Barrow Edge and her Emergency Conservation Committee bought and leased land on the ridges and actively discouraged hunters, eventually making the area protected property. There are now five National Natural Landmarks within the Kittatinny-Shawangunk migration corridor, which crosses over New York, New Jersey, and Pennsylvania, providing watch sites in the mountains to which thousands of birders make their own annual migration in order to take in the spectacle.

As the twentieth century progressed, good field guides were published and affordable binoculars and camera equipment became generally available. Bird-watching evolved into an accessory-laden sport and a hobby that remains universally popular.

Question 2: Why are people fascinated by raptors?

Answer: The human fascination with birds of prey may have begun with falconry. "There is no practical reason for falconry to exist, although it probably originated as long as four thousand years ago," according to falconry enthusiast Stephen Bodio, "probably as a recreation for the lucky few rather than a practical way to put meat on the table." A falconer's bond with his or her bird is unique; some go so far as to describe it as a religion or an addiction. Stories are told about marital strife when a falconer who is "manning" a new bird (see chapter 7, question 7: How are raptors trained?) spends family time with the bird on the glove—inside and outside of the house.

Raptors evoke intense emotions in their admirers, even if they are only observers. The frisson that attends spotting a raptor in the wild has a different quality than the pleasure one feels spotting a hummingbird. The strong reaction aroused by birds of prey may result from a powerful mix of conflicting emotions. On the one hand, we may experience what Harvard biologist E. O. Wilson calls "biophilia," an instinctive, subconscious desire on the part of human beings for a connection with other living organisms. This feeling is most acute when we respond to a human baby, especially our own, but it is also a common response when we see animals. For example, a pet store with puppies in the window almost always draws a small crowd, oohing and aahing at their antics. The evolutionary basis for biophilia, in Wilson's view, may be that other species are our kin, since all life is descended from a single ancestral population.

Audubon's Iceland or Jer Falcon

John James Audubon was an iconic painter of birds. He published his four-volume elephant folio *The Birds of America* between 1827 and 1838, comprised of 435 hand-colored etchings. He had spent a lifetime observing birds in the still-unspoiled expanses of North America, rendering them life-sized in detailed watercolors.

Audubon was particularly fascinated by birds of prey and spent countless hours observing their habits in the wild. In his elephant folio, he devoted twenty-seven plates to them, and in his five-volume *Ornithological Biography* designed to accompany the folio volumes, he drew on his field experience to paint word pictures of the birds for his readers.

Although Audubon used both preserved skins of birds and recently dead specimens as models, he preferred to work from live specimens and kept many birds of prey in captivity at one time or another in order to draw them from life. This practice, combined with his many observations of the behavior of wild individuals, allowed him to combine details of plumage with poses that brought a sense of strength and activity to his finished watercolors.

As volume 3 of his elephant folio was being readied for publication, Audubon wanted to add illustrations of some of the more northern species. So, in 1833, he chartered a schooner and with his son John Woodhouse Audubon and some young friends, he sailed to Labrador. His stay was late in the short summer season, and the weather was almost continually cold and stormy. His journal reflects a trip filled with daily hardships and disappointments, as in this entry: "Another horrid, stormy day. The very fishermen complain." While his son and the younger members of the party did most of the collecting of specimens, Audubon spent most of his time observing and drawing. A special area on the vessel that had been prepared to give him good light and to keep his drawing materials dry did neither. Only his dogged determination kept him at his self-imposed task.

Audubon's Iceland or Jer Falcon

One of the bright spots of the expedition was his first view of what he called the "Iceland" or "Jer Falcon" *Falco islandicus,* now called the Gyrfalcon *F. rusticolus.* He saw only the dark phase of the bird in Labrador and spent a lot of time observing it. He noted in his journal that individuals flew much like a Peregrine Falcon *F. peregrinus* and resembled it "much in form, but neither in size nor color. Sometimes they hover almost high in air like a small Sparrow Hawk when watching some object fit for prey on the ground, and now and then cry much like the latter, but louder in proportion with the difference of size in the two species." His younger companions finally managed, almost at the end of their stay, to collect two specimens—a male and a female. The exhausted Audubon immediately set to work drawing these individuals before the colors of their fleshy parts faded. His drawing was eventually published as Plate 196 in the elephant folio (see figure 30).

But Audubon was not completely satisfied with his Iceland Falcon because he had not managed to find a white-phase individual. On a trip to England, he met John Heppenstall of Sheffield, an acquaintance of his son who had such a bird in captivity. In 1837, when the Audubons were again living in England, Heppenstall offered to send it alive to Audubon so that he could draw it. Although the bird died of an "affectation to the oesophogus" before Heppenstall could send it, he sent the body, which arrived in good condition. Audubon used this specimen as the model for two birds in elephant folio Plate 366 (see figure 31), and gave the internal organs to William Macgillivray, curator of the Museum of the Royal College of Surgeons of Edinburgh, which he described and depicted in the *Ornithological Biography.*

—Mary LeCroy, research associate, Department of Ornithology, American Museum of Natural History

Figure 30. A pair of dark morph Gyrfalcons *Falco rusticolus*, painted by John
James Audubon in the nineteenth century. Gyrfalcons are closely related
to Saker Falcons *Falco cherrug*, native to some of the harshest deserts in the
world. *(Photo courtesy of the American Museum of Natural History)*

Figure 31. Largest of the true falcons, the magnificent Gyrfalcon *Falco rustico-lus*, here in a painting by John James Audubon, is circumpolar in distribution and exhibits several color variations or morphs. These are white Gyrfalcons, highly prized by falconers throughout the ages. *(Photo courtesy of the Audubon Society)*

Sky Burial

In some parts of the world, it is believed that vultures provide a natural way of recycling the dead. Where there has been a severe decline in the vulture population, these practices have been curtailed, but they still exist. Buddhists in Northern India and Tibet have traditionally put out cadavers for Himalayan Griffons *Gyps himalayensis*, Cinereous Vultures *Aegypius monachus*, and Bearded Vultures *Gypaetus barbatus*. Niema Ash and Pamela Logan, authorities on Tibet, have each witnessed this ritual and describe it similarly. According to their reports, in preparation for what is known as a sky burial, the corpse is cleaned, wrapped in a white cloth, and transported to an excarnation site near a monastery. It is placed in the fetal position within a large circle of stones to mirror the beginning of life, and the local monks ceremonially present the body to the vultures. They make short work of the body, and the bones are eventually reduced to dust.

Zoroastrians also have historically used this type of funerary practice to dispose of their dead, and some Parsees in India and elsewhere still hold to this ancient tradition. Their dead have been offered to Red-headed Vultures *Sarcogyps calvus* and Indian White-backed Vultures *G.bengalensis*. The preparation is similar, and the body is placed in a round, stone bowl-like structure almost 100 yards in circumference (91 meters) and almost 40 yards high (36 meters). This practice reflects a belief that the dead should contaminate neither the living nor the earth.

On the other hand, we may also experience biophobia when we see a raptor, heightened anxiety and excitement at being near a predator that kills and ravages its prey. Predators of all sorts threatened the survival of early humans, and an adaptive biophobic fear/avoidance response to the dangers they posed gave them (and gives us) an evolutionary advantage. That we are both fascinated and attracted by the powerful life force exhibited by birds of prey and simultaneously cringe and feel the urge to flee the bloodletting that they symbolize helps explain the singular appeal that raptors exert upon us.

Question 3: Where can I see raptors?

Answer: Some rehabilitation facilities offer educational demonstrations, either on-site or at venues to which they have been invited. The birds are carried on a glove to open-air perches, to which they are tethered with soft leather jesses, and they sit quietly as they would in the wild. Raptors naturally sit quietly to conserve energy, unless they need to hunt or are searching for a mate. Attending a demonstration offers the unusual opportunity to inspect these beautiful birds almost face-to-face, but from a safe distance. Birds are usually flown from one gloved staff member to another, motivated by a morsel of food as a reward. Along with the demonstration, the staff tells the observers about the birds, their habits, and how they came to the rehabilitation facility.

There are a variety of ways you can observe and learn more about raptors in the wild. You should be able to see some common species of raptors near where you live, even if you live in a city. Check with your local birding group for suggestions about where to look. There are special sites where large numbers of migratory birds can be seen at certain times of the year, and some groups offer raptor-watching trips and tours. Appendix A lists some popular places where you can see raptors, and Donald Heintzelman's *Guide to Hawk Watching in North America* contains an extensive list of places to see wild raptors in Canada and the United States, organized by province and state (see appendix B).

It is helpful to take along an identification guide for the birds in the area, so you can identify the raptors you see and avoid confusing them with birds that may have similar features. The plumage of some birds changes as they reach adulthood, a process that can take from one to five years depending on the species. A truly useful guide should illustrate the juvenile and adult plumage, as well as the different appearance of males and females. Take along binoculars or a spotting scope on a tripod if possible, since you may find yourself watching birds that are sitting in one spot. You can enjoy rich details if you set up a scope from a respectful distance.

From about the beginning of August to year's end, birdwatchers track more than 150 species of songbirds and 16 species of raptors as they migrate, moving south from their summer ranges in the woodlands of the United States and Canada to as far away as Central and South America in search of a dependable supply of food. Avid observers can spot hawks, eagles, Ospreys *Pandion halietus,* falcons, vultures, and other raptors that soar effortlessly overhead. In some areas, such as the Hawk Mountain Sanctuary in Pennsylvania, the terrain and the air currents cause the birds to float within a few feet of the rocks on which the watchers sit. In addition to recreational birdwatchers, volunteers conduct scientific counts, logging in more than twenty thousand birds each year as they pass over this area. The annual migration counts are used to monitor changes in the regional populations of the birds, and the Sanctuary is considered one of the most important institutions in the world for raptor protection and research. Eliat, at the southernmost tip of Israel, is another ideal observatory and a research center for studying raptor migration because of its location (see chapter 3, question 6: How do raptors find their way during migration?).

Question 4: What should I do if I find an injured or dead raptor?

Answer: It is important to handle an injured or dead bird as little as possible, and always to wear gloves to avoid getting "footed" by the bird's sharp talons as it may instinctively try to defend itself. A raptor that has been hit by a car and is unconscious might be easy to handle, but an alert injured bird can be dangerous. Call your local state fish and wildlife agency, animal control officer, local rehabilitator, or veterinarian to report what you have found and to get directions as to what to do next.

Rehabilitating a Golden Eagle

A Golden Eagle *Aquila chrysaetos*, later named Canyon, was found lying at the side of a road near a small airstrip in Moab, Utah, early in 1995. Examination revealed that he was an adult bird at least five years old, apparently in shock. Within a day of being taken into the care of a rehabilitator, he lost all his primary feathers. The veterinarian who examined him found no broken bones, but there was permanent nerve damage to one wing. It seemed probable that he had gotten caught up in the wash of a plane, lost control, and landed on his wing. It was clear that he would never be able to fly well enough to survive in the wild.

After several months in a rehabilitation facility in Utah, Canyon arrived at Hawk Creek Wildlife Center in East Aurora, New York, where he spent some time in quarantine. He

(continued)

Figure 32. Golden Eagles *Aquila chrysaetos* get their common name from the beautiful golden-colored feathers at the nape of the neck. They are large and powerful booted eagles, meaning that feathers cover the entire leg down to the toes. *(Photo courtesy of Loretta Jones)*

then moved into an enclosure with One Wing, a resident Bald Eagle *Haliaeetus leucocephalus,* and eight years later, Cherokee, a female Golden Eagle, arrived as a companion animal, and they now share an enclosure.

Hawk Creek Center, founded in 1987 and funded by private donations, is a licensed facility that rehabilitates injured and orphaned wildlife and releases them into protected environments whenever possible, taking in approximately five hundred patients a year. Animals that cannot be released are assessed for educational use or breeding programs, and if suitable, they are placed at facilities across the country.

Many hours of training by volunteers at Hawk Creek have made Canyon comfortable traveling and riding on his trainer's arm in front of a large crowd. He has participated in educational programs in New York and surrounding states, giving visitors the unique experience of standing within a few feet of a huge, powerful bird that usually resides in remote mountains and can be seen only from afar.

Golden Eagles are large, impressive birds that tend to mate for life, and their life span in the wild is believed to be thirty years or longer. They have an average wingspan of more than 6 feet (2 meters), a body length that can exceed 3 feet (1 meter), and an adult weight of 7 to 14 pounds (3.2 to 6.4 kilograms). They are able to catch prey as large as a small mountain sheep. Although they are not federally listed as endangered, Golden Eagles are considered endangered in New York State and have been protected in the United States since 1963. They are fairly common in the western states, Alaska, and western Canada, and can also be found in Eurasia and Siberia.

Used for falconry in many European countries, Golden Eagles are popular in Germany, where they are preferred for catching large European hares. Their typical prey includes rodents, birds, waterfowl, and reptiles, as well as carrion. Golden Eagles have a grip strength of 700 pounds per square inch per talon, making them one of the most powerful eagles in

Rehabilitating a Golden Eagle

the world. One falconer (Stephen Bodio) reported that a zoo specimen with concrete-dulled claws was still strong enough to crack a bone in his hand through a too-light glove, but a Golden Eagle can crush a bone in a falconer's hand even in a heavy glove. Few Americans attempt to train them, primarily because of their protected status, but also because of their size and strength and their capacity to go after large prey. They need to be handled by an experienced falconer with lots of hunting land, so they do not capture someone's pet or injure their handler.

Remember that it is illegal to possess a raptor without a permit, so proceed carefully.

If the bird is dead and not unmanageably large, you may be directed to place it in a plastic bag marked with the date and the location where the bird was found and to put it in a freezer or refrigerator so that the cause of death may be determined. The cold also preserves the specimen so that it may be suitable for donation to a museum. If the bird is alive but injured or appears ill, prepare a cardboard box large enough so that the bird will not be crowded. Punch some holes in the top and sides of the box and line the bottom with a towel or an old t-shirt (so it is not slippery). You can use a pet carrier instead of a box, if it is large enough, as long as you cover the entry door with a cloth or towel. Then throw a jacket or blanket over the bird and carefully lift it in the blanket, place it in the container, and close it securely. Do not attempt to feed the bird or give it water. Just keep it warm, dark, and quiet until you get instructions as to where to take it.

Question 5: What attracts raptors to live in cities?

Answer: Coauthors Carol Butler and Peter Capainolo live and work in New York City, and we will use our city as an example of what occurs in many places around the world. Raptors and other birds live in and around a city like New York in part because they are naturally drawn to the high ledges on the tall buildings. New York, like other cities and states, makes an effort to attract and retain bird populations by creating green spaces, landscaping, and lighting systems that reduce collisions between migrating birds and buildings.

Among the efforts to help affected populations rebound once DDT was made illegal in the United States, scientists sought to settle captive-bred falcons in urban areas. Wild Peregrine Falcons *Falco peregrinus* had once nested on tall buildings and bridges with access to an abundant population of Rock Pigeons *Columba livia,* so scientists provide gravel-filled nesting boxes in some of those locations and peregrines usually accept them and settle in to breed. In recent years, peregrines have been reestablished successfully in the United States in New York, Boston, Baltimore, Pittsburgh, Detroit, Cleveland, Columbus, Buffalo, Rochester, and scores of cities in the Midwest, West, and especially along the West Coast, as well as in London, and in Toronto, Hamilton, and Ottawa, Canada.

Prey animals and raptors become part of the cities' fabric in parks and other areas of uninhabited land. Brooklyn's 585-acre Prospect Park, Manhattan's 843-acre Central Park, and the Bronx's 2,766-acre Pelham Bay Park are favorite venues in New York City for birding and nature walks of all types. City residents help make urban areas hospitable to animals by planting private and community gardens, and by landscaping around their homes to bring the flavor of the countryside into the cityscape. Many local residents are active members of national and local bird clubs in the city and in surrounding and upstate counties. For example, the New York State Ornithological Association, founded in 1946, lists more than fifty member clubs on its Web site. These groups are involved in counting and reporting local species in large and small parks and on the streets, and they are energetic advocates for city raptors and other urban wildlife.

Many states have Rare Bird Alert phone numbers—New York State has seven, including a special one for New York City. Word

Figure 33. Northern Saw-whet Owl *Aegolius acadicus* well camouflaged in a tree in New York's Pelham Bay Park in the Bronx. *(Photo courtesy of Carol A. Butler)*

spreads quickly when an unusual avian visitor appears anywhere in the area. When the police were called to Fordham University to rescue one of the campus's Red-tailed Hawks *Buteo jamaicensis* that was injured, it was reported in the local newspaper. When Peregrine Falcon hatchlings were seen in a nesting box on top of the 693-foot (211 meter) tower of New York City's Verrazano-Narrows Bridge, the news quickly spread. And when five Eastern Screech Owls *Megascops asio* were recently released in New York's Central Park, the event was photographed and publicized.

A 2007 bird count reported thirty-two breeding pairs of Red-tailed Hawks nesting in New York City, and many more have been observed in the city by its devoted cadre of birdwatchers (see also chapter 1, question 2: Where in the world are raptors found?). Cities like New York make an effort to attract and retain birds and other wildlife with projects like the Watershed

Protection Program. Under the auspices of the city's Department of Environmental Protection, the phase of the program that ended in 2007 protected approximately seventy-five thousand acres through purchase or easement, and the effort is projected to continue. Its goals are to help prevent habitat fragmentation, to work to assure the survival of wildlife, and to encourage outdoor activities on city- and state-owned lands. Protecting such a large area also helps maintain water quality in the trout streams and reservoirs of the watershed that supplies New York City's drinking water.

Question 6: Of what value are raptors to the environment?

Answer: Raptors are generally an environmental boon because they prey on rodents, other small mammals, and insects of many species that are agricultural pests. A family of Barn Owls *Tyto alba* can eat thousands of mice and other small mammals in a year. For this reason, increasing numbers of farmers are preserving trees suitable for nesting and setting out nest boxes to encourage raptors to remain on their property.

Many federal, state, and local laws protect raptors from being shot, but changes in their habitat and the subsequent drop in the population of prey animals has diminished raptor populations. They are apex predators, feeding at the top of the food web, and their presence or absence indicates the condition of the habitat because they are attracted by healthy prey populations for their food supply.

NINE

Research and Conservation

Question 1: Why do we need to study raptors?

Answer: The geographic range of a species is a reflection of how births, deaths, and movement respond to environmental variations over time, say Robert Holt and Michael Barfield of the University of Florida. Each species maintains its niche in the food web either by managing to stay alive, or by reproducing copiously enough to offset the mortality caused by natural enemies. Understanding the environmental causes that limit the range of a species and the consequences of these limits are key research issues in ecology and evolutionary biology. Studying the role played by raptors, apex predators at the top of the food web, is instrumental in understanding the total picture.

Brothers Frank and John Craighead, in a landmark field study published in 1956, set out to explore the dynamics and ecology of raptor predation. They conducted an exhaustive quantitative analysis of raptor and prey activity over two years in Superior Township in Michigan, motivated by their belief that "predation can be understood and evaluated by determining the effect of the aggregate population of raptors on the aggregate population of prey over an extended period of time." All subsequent raptor literature and research builds on their study.

Among this research is a study by Peter Banks of the University of Turku, Finland, and colleagues that describes how predation can sometimes prevent the extinction of a prey species. Prey is captured in proportion to its availability and vulnerability, and raptors prey on and help control the populations of some of the most abundant animals in the world: insects, rodents, and other small animals that are agricultural pests. On predator-free islands in the Baltic Sea, Banks and his colleagues found the mushrooming population of small rodents (field voles *Micro-tus agrestis* and bank voles *Clethrionomys glareolus*) overgrazed the plants that are their food supply to the point that the vole population crashed and bordered on extinction. On islands where predators were present, the predators hunted and killed significant numbers of voles, which kept the voles from overrunning their habitat; their population on those islands remained stable and robust.

In another exploration of habitat use, Sergei Surmach and his colleague Jonathan Slaght are conducting an ongoing study of the endangered Blakiston's Fish Owl *Ketupa blakistoni* in Russia. Focusing on long-term conservation issues in a cold, remote area where resources are being depleted by logging, they capture the owls briefly to fit them with GPS dataloggers, which they attach like small backpacks with Teflon-coated cotton straps. The dataloggers record an owl's location every eleven hours for about a year. When they recapture the owl the following year, the downloaded data show the locations where the owl has been, key to understanding how the owls use their habitat.

In that habitat, where temperatures drop to minus 40 degrees Fahrenheit in the winter (minus 40 degrees Celsius), the owls live in pairs and search for openings in the frozen rivers to prey on fish. Blakiston's Fish Owl, family Strigidae, may be the largest owl in the world, standing almost 2.5 feet tall and weighing 9 pounds or more. They sometimes maintain more than one nest tree, perhaps because strong winds can easily topple the decaying trees in which they nest. They usually raise only one nestling every other year in this harsh habitat.

In 1968, ornithologist Leslie Brown, formerly of the Colonial Agriculture Service, Kenya, and Dean Amadon, at the American Museum of Natural History for most of his career, published *Eagles, Hawks, and Falcons of the World*. Their aim was "to bring together in one work all the available knowledge on

Striated Caracaras

Striated Caracaras *Phalcoboenus australis,* the world's southernmost birds of prey, are among the rarest and most unusual. The ten caracara species comprise a fascinating and odd tribe of falcon relatives, found in nearly every kind of habitat from southern Texas to Tierra del Fuego. They fill ecological niches usually occupied in the rest of the world by crows. Striated Caracaras live and breed only on the outer Falkland Islands and on scattered, exposed islands south and west of Tierra del Fuego, including Cape Horn and the tiny archipelago of Diego Ramirez north of the Antarctic Peninsula. The southernmost point on the South American continent, Islote Aguila (Eagle Islet), may take its name from the Striated Caracaras that breed there.

Jonathan Meiburg, an authority on these birds, is our source for this material. His fieldwork was done in the outermost islands of the Falkland archipelago and on Isla de los Estados in Argentina. He observed that Striated Caracaras feed on colonial seabirds that breed on the remote islands during the austral summer (summer in the southern hemisphere), including penguins, albatrosses, shags, and small burrowing petrels. The eggs, nestlings, and adults massed at these colonies are a concentrated food resource for Striated Caracaras, although they also scavenge on carrion, invertebrates, shellfish, and the excrement of seals and sea lions. Although they are adept at using their long tails as rudders while soaring on the constant sea winds, the caracaras also move with ease on the ground. Using their long, strong legs and large feet, they can crack the tough shell of an albatross egg or pull the head off a young penguin. They appear to have excellent night vision that they use to prey on seabirds returning to their burrows after sunset.

Steeple Jason, a 790-hectare island in the Falklands that supports breeding colonies of Black-browed Albatrosses *Thalassarche melanophrys,* Rockhopper Penguins *Eudyptes chrysocome,*

(continued)

Striated Caracaras, *continued*

Giant Petrels *Macronectes halli,* Wilson's Storm Petrels *Oceanites oceanicus,* King Cormorants *Phalacrocorax albiventer,* and other seabirds, was also home to seventy-two breeding pairs of Striated Caracaras in 1997. The caracaras have adapted to the seabirds' spatial concentration by defending territories at the colonies' periphery that can be separated by as little as fifty meters.

During the austral winter when the seabird colonies are largely vacant, Striated Caracaras face a relatively lean period. They cannot forage at sea and do not appear to migrate. This deprivation may have exacerbated their attraction to novel or unfamiliar objects, a trait so pronounced that it is the source of their frequent characterization as mischievous or cheeky. "From their point of view," observed James Hamilton, government naturalist in the Falklands from 1919 to 1949, "every strange object requires immediate examination; . . . one bird with which I had a slight acquaintance would play for a long time with an empty sardine tin."

Charles Darwin, who was intrigued by the birds during his visits to the Falklands in 1833 and 1834, called them "tame and inquisitive, quarrelsome and passionate." Captain Charles Barnard, a seal hunter who was marooned in the Falklands in 1812, wrote with regard to the caracaras that "the sailors who visit these islands, being often much vexed at their predatory tricks, have bestowed different names on them, characteristic of their nature, as flying monkeys, flying devils, etc., etc. I have known these birds to fly away with caps, mittens, stockings, powder horns, knives, steels, tin pots, in fact everything which their great strength is equal to; . . . they compel us to secure our provisions, by covering them with the sails of the boat, which we fastened down by stones, and then direct the dog to lie down by them to prevent these harpies from hauling off the stones and sails."

These behaviors did not endear Striated Caracaras to the sheep farmers who arrived in the Falklands in the late nine-

Striated Caracaras

teenth century and saw the birds as a threat to their livestock. The government of the islands placed a bounty on the "Johnny Rooks" for several years, and even after the bounty was suspended, some farmers continued to shoot the unwary birds, which, as James Hamilton wrote, "had not learnt that man is dangerous." This persecution, combined with the transformation of Falkland habitats by grazing, fire, and the introduction of non-native predators like rats and cats, reduced the suitable habitat for the caracaras to a fraction of its former extent. Striated Caracaras are now protected by law, but their population appears to have stabilized in recent years at about five hundred breeding pairs, and they may have saturated most of the available habitat.

The species as a whole faces an uncertain future. Many of the seabird species on which Striated Caracaras depend are threatened or endangered. The International Union for Conservation of Nature currently lists Striated Caracaras as "near-threatened" due to their small population size, estimated roughly at a thousand breeding pairs. Resourceful as they are, the survival of these rare, curious raptors may be tied to the fate of their ocean-going prey and to the health of the world's marine ecosystems.

the diurnal birds of prey in the world." It is the consensus that they succeeded, and a great deal of environmental policy and research flowed from the massive amount of information they made available.

Why do populations disperse or crash? What are the environmental limitations if there are no apparent terrestrial limits? The studies described here, as well as other ongoing research, examine the overt and subtle environmental influences on range limits and on the physiological tolerances and constraints that species experience. A 2009 report, based on information gathered by the U.S. Fish and Wildlife Service and several other conservation-oriented organizations, found that "habitat destruction, pollution, and other problems have left nearly a third of the nation's 800 bird species endangered, threatened, or in serious decline." The purpose of raptor research in the twenty-first century is ultimately to stem the loss of our natural resources.

Question 2: How do museums accumulate their collections of specimens?

Answer: Natural history museums serve as research and educational institutions, and many of them have a long history of collecting plant and animal specimens. Before the advent of wildlife management and protection laws, hunters, amateurs, and professional scientific collectors were free to trap, net, and shoot any animals without limit. Professional ornithologists collected many raptor species for museums and prepared them as study skins, cotton-stuffed skins that fit into special specimen cases. Each specimen is labeled with valuable information such as the date and location the specimen was obtained, the condition of its sex organs, and its fat content (important in relation to migration and quality of life in the habitat). Eggs were also collected and blown (the contents removed), and relevant data were recorded. The importance of scientific specimens cannot be overemphasized. Museum collections of pre-DDT Peregrine Falcon *Falco peregrinus* eggs, for example, were the only resource for determining the degree of eggshell thinning caused by the pesticide when it almost wiped out populations of this magnificent raptor in the mid-1950s.

Today, under special permits, museums may acquire a limited number of raptor specimens by collecting them in the field. The vast majority of new museum specimens, however, are donated by raptor rehabilitators, zoos, and falconers and are birds that have succumbed to injury or disease or were found dead. Many people object to collecting birds for scientific purposes, but the number of specimens collected today is minimal compared to all the other dangers animals face in the wild. Including some of the first specimens that were collected in the 1600s and 1700s, all the specimens in all the museums of the world number fewer than the ducks hunters shoot in just one year in North America.

Question 3: How are raptors captured for study?

Answer: Researchers capture raptors for study using methods that falconers have employed for centuries. Examining, banding, or removing a nestling falcon or Golden Eagle *Aquila chrysaetos* usually requires rock climbing and rappelling skills, as these birds breed on cliff ledges hundreds of feet from the ground. Care must be taken so nestlings that cannot fly do not scurry away from the ledge and fall.

Many raptor species build stick nests in a crotch high in a tree, or lay their eggs in holes or crevices of dead trees. To reach the young of these species requires climbing irons affixed to one's boots, long iron spikes that can be thrust into the tree trunk in a walking motion. A thick belt that encompasses both climber and trunk is slid up the trunk with every step. This kind of climbing takes strength, coordination, and courage, especially if an irate parent raptor is diving at the climber, defending its young. An aluminum telescopic ladder may be enough to reach more accessible nests.

Once a nest tree has been climbed or disturbed in any way, a wide strip of aluminum flashing must be nailed completely around the trunk several feet from the ground and spray-painted green and brown to blend with the bark. Raccoons *Procyon lotor* and other predators that follow human scent to the tree run into this wide, slippery metal girdle, which prevents them from gaining a foothold, climbing to the nest, and preying on the eggs or young.

There are two main methods for capturing adult raptors and immature birds that are fully fledged and hunting on their own. One is the use of the Bal Chatri trap (BC), a square cage of the type of heavy metal screen typically used for a fence or rabbit hutch. Small, lightweight BCs are used to trap small raptors, larger and heavier BCs with thicker screen for larger ones. Usually the bottom of the BC is a square slab of wood, with the wire cage making up the other five sides, but there are many different styles.

A small door allows a live mouse, rat, sparrow, or pigeon to be placed inside as bait. The wire cage is festooned with monofilament slip nooses. When a perched raptor is sighted, the BC can be placed on the ground within its line of sight. If hungry, the raptor will swoop down to capture the bait animal and will get its toes tangled in the slip nooses. Because the BC is weighted and too heavy for the hawk to lift, the trapper can then approach and gently remove the nooses from the hawk's toes.

The second common way to trap raptors is to set up a trapping station along a fall migration route where abundant passage birds are making their first flight from their breeding grounds. A blind made of wood or canvas provides camouflage for the trappers inside. A half circle of mist nets—fine dark nets that are almost invisible—is set up several yards from the blind, and a bait pigeon or starling is tethered between the nets and the blind with a line that runs into the blind. When a migrating raptor is sighted, the trapper gives the bait bird a gentle tug with the line, and the hawk perceives it as injured quarry. The raptor comes in low for the kill but is caught in the mist nets before reaching the bait.

There are modifications of these two methods depending on species, terrain, and time of year, but trapping has changed little over time.

Question 4: How are raptor skins prepared for study or exhibit?

Answer: Properly prepared and preserved animal skins are used for all sorts of research, as well as for exhibits and artistic renderings. Specimens are salvaged from rehabilitators or zoos when the animals die, or collected from the wild with the proper permits. The date and source of acquisition are noted, along with the details of pattern and coloration. The label remains attached to every animal in an orderly collection, and proves invaluable when present and future generations of researchers need to refer to the specimen. For example, Paolo Galeotti of the Universita degli Studi di Pavia, Italy, and colleagues recently conducted a study in which they examined the color of the plumage for museum study skins of Scops Owls *Otus scops* that had been collected and acquired between 1870 and 2007. They correlated the observed color variations of skins acquired at different times with records of temperature and rainfall.

To prepare a study skin that will be useful to scientists, an incision is made in the abdomen of the bird and the skin is removed from the muscles. The feathers are undisturbed because they are rooted in the skin like hairs or fur. The skull, wing bones, and leg bones, which also remain attached to the skin, are cleaned, and all remaining meat and fat is removed. Cotton is placed in the eye sockets, and a wooden dowel wrapped with cotton is placed inside the skin to give the bird shape. Then the incision is stitched closed, and the bird is preserved by drying, either wrapped in cotton or pinned to a frame so that it retains its shape.

Question 5: Are any raptors endangered?

Answer: Raptors listed as endangered include the California Condor *Gymnogyps californianus*, Snail Kite *Rostrhamus sociabilis*, Hawaiian Hawk *Buteo solitarius*, and Aplomado Falcon *Falco femoralis*. Raptors listed as threatened include the Bald Eagle *Haliaeetus leucocephalus*, Crested Caracara *Caracara cheriway*, Northern Spotted Owl *Strix occidentalis caurina*, and Mexican Spotted Owl *S. o. lucida*.

The good news is that raptors such as Cooper's Hawks *Accipiter cooperii*, Red-shouldered Hawks *B. lineatus*, Merlins *F. columbarius*, Ospreys *Pandion haliaetus*, and Bald Eagles have responded positively to the banning of pesticides and to the stringent regulation of hunting: their populations have rebounded.

No one knows exactly how many Bald Eagles there are in North America. Estimates range as high as seventy-five thousand, most of them in Canada, which has its own regulations regarding eagles. In the lower forty-eight states, there are approximately eight thousand Bald Eagles, reflecting the rebound from near extinction caused by toxic chemicals, loss of habitat due to urbanization, and poaching. Eagles have been removed from the endangered species list, but they are protected under the Migratory Bird Treaty Act and the Bald and Golden Eagle Protection Act.

Because Native Americans have traditionally used eagle feathers for ceremonial and religious purposes, the National Eagle Repository was established in the 1970s to make feathers and parts of Golden and Bald Eagles available for religious purposes to people who apply for a permit and provide certification of tribal enrollment from either the Bureau of Indian Affairs or the Tribal Enrollment Office. According to the National Eagle Repository Web site, about five thousand people are on the list of tribal members approved to receive an eagle or eagle parts, for which the wait is three and a half years.

About a thousand dead eagles are shipped each year in the United States to the National Eagle Repository at the Rocky Mountain Arsenal National Wildlife Refuge in Denver, Colorado. Each bird is assigned a number, carefully studied, and its details recorded. The bird is then frozen until it can be shipped to the next person on the waiting list, who then has to go to the end of the line to receive any part of another eagle.

APPENDIX A

Places to See Birds of Prey

Cape May National Wildlife Refuge
24 Kimbles Beach Road
Cape May Court House,
New Jersey 08210
www.fws.gov/northeast/capemay/
Tel: 609-463-0994
E-mail: howard_schlegel@fws.gov

Fisherman Island National Wildlife Refuge (part of the Virginia
    Barrier Island Chain, Chesapeake Bay)
Eastern Shore of Virginia National Wildlife Refuge
5003 Hallett Circle
Cape Charles, Virginia 23310
www.fws.gov/northeast/easternshore
Tel: 757-331-2760
E-mail: fw5rw_esvnwr@fws.gov

Hog Island Audubon Center/Maine Audubon
20 Gilsland Farm Road
Falmouth, Maine 04105
www.maineaudubon.org
Tel: 207-781-2330

New York's Central Park
Central Park Conservancy
14 East 60th Street
New York, New York 10022
www.centralparknyc.org/site/Search?query=bird
Tel: 212-310-6600

Point Pelee National Park of Canada
407 Monarch Lane, RR 1
Leamington, Ontario
Canada N8H 3V4
www.pc.gc.ca/pn-np/on/pelee/index_E.asp
Tel: 519-322-2365 or 888-773-8888
E-mail: pelee.info@pc.gc.ca

Point Reyes National Seashore
1 Bear Valley Rd.
Point Reyes Station, California 94956
www.nps.gov/pore/contacts.htm
Tel: 415-464-5100, ext. 2

Rock Creek Park
3545 Williamsburg Lane, NW
Washington, D.C. 20008
www.nps.gov/rocr/naturescience/index.htm
Tel: 202-895-6070

Roy P. Drachman Agua Caliente Regional Park
12325 East Roger Road
Tucson, Arizona 85749
www.pima.gov/nrpr/parks/agua_caliente/index.htm
Tel: 520-749-3718
E-mail: AguaCalientePark@pima.gov

Santa Ana National Wildlife Refuge
Jodi Stroklund, Refuge Manager
Route 2, Box 202A
Alamo, Texas 78516
www.fws.gov/southwest/refuges/texas/santana.html
Tel: 956-784-7500

Sauvie Island
Sauvie Island Wildlife Area
18330 NW Sauvie Island Road
Portland, Oregon 97231
www.dfw.state.or.us/resources/visitors/sauvie_island_wildlife_area
    .asp
Tel: 503-621-3488

Whitefish Point Bird Observatory
16914 N. Whitefish Point Road
Paradise, Michigan 49768
www.wpbo.org/
Tel: 906-492-3596

Note: For a more comprehensive list and detailed recommendations, see Donald S. Heintzelman, *A Guide to Hawk Watching in North America* (Helena, Mont.: Falcon Press, 2004).

APPENDIX B

Recommended Reading and Web Sites

Books

Berger, C. 2005. *Owls: Wild Guide*. Mechanicsburg, Pa.: Stackpole Books.

Bildstein, K. L. 2006. *Migrating Raptors of the World: Their Ecology and Conservation*. Sacramento, Calif.: Comstock Publishing.

Bodio, S. 1992. *A Rage for Falcons*. Boulder, Colo.: Pruett Publishing.

Bosakowski, T., and D. G. Smith. 2002. *Raptors of the Pacific Northwest*. Portland, Ore.: Frank Amato Publications.

Clark, W. S. 2000. *A Field Guide to the Raptors of Europe, the Middle East, and North Africa*. New York: Oxford University Press.

Dunne, P., D. Sibley, and C. S. Dunne. 1988. *Hawks in Flight: The Flight Identification of North American Migrant Raptors*. Orlando, Fla.: Houghton Mifflin.

Ferguson-Lees, J., and D. A. Christie. 2006. *Raptors of the World: Princeton Field Guides*. Princeton, N.J.: Princeton University Press.

Gallagher, T. 2008. *Falcon Fever*. Orlando, Fla.: Houghton Mifflin.

Heintzelman, D. S. 2004. *Guide to Hawk Watching in North America*. Helena, Mont.: Falcon Press.

Liguori, J. 2005. *Hawks from Every Angle: How to Identify Raptors in Flight*. Princeton, N.J.: Princeton University Press.

Peeters, H., and P. Peeters. 2005. *Raptors of California: California Natural History Guides*. Berkeley, Calif.: University of California Press.

Weidensaul, S. 2007. *Of a Feather*. Wilmington, Mass.: Harcourt.

Wheeler, B. K., and W. S. Clark. 2003. *A Photographic Guide to North American Raptors*. Princeton, N.J.: Princeton University Press.

Wheeler, B. K. 2007. *Raptors of Eastern North America: The Wheeler Guides*. Princeton, N.J.: Princeton University Press.

Web Sites

Archives of falconry. Peregrine Fund, http://www.peregrinefund.org/
american_falconry.asp#heritage%3E.
The Archives of Falconry is the only library and museum in the
world wholly dedicated to preserving the literature, art, equipment,
personal journals, and photo albums of falconers from all over the
globe. It is part of the Peregrine Fund of the World Center for Birds
of Prey in Boise, Idaho.

Dobney, Keith. Ancient falconry. FirstScience.com, http://www.first
science.com/site/articles/dobney.asp.

Falconry, ecology, education. The Modern Apprentice, http://www.the
modernapprentice.com/history.htm.

Nick Dunlop photography (raptor photographs), http://www.nick
dunlop.com.

Richard Ettlinger nature photography (raptor photographs), http://
www.richardettlinger.com.

Vadim's works (raptor paintings by Vadim Gorbatov), USA Raptor Education Foundation, http://www.usaref.org/VadimWorks.htm.

Webcam with live video in the nest of an eagle with three chicks, zaplive
.tv, http://www.zaplive.tv/web/hwf-sidney2.

Windsor Nature Discovery, wildlife identification posters, http://www
.nature-discovery.com.

APPENDIX C

Raptor Species Mentioned in the Book

Common Name	Scientific Name	Range
African Black Sparrowhawk	*Accipiter melanoleucus*	Senegal and Gambia east to Gabon, Congo, and Central African Republic; East Sudan and north and west Ethiopia; Gabon and Zaire east to Kenya and south to Angola and South Africa; Pemba and Zanzibar.
African Fish Eagle	*Haliaeetus vocifer*	Most of continental Africa south of the southernmost edge of the Sahara Desert.
African Harrier-Hawk	*Polyboroides typus*	Most of Africa south of the Sahara.
African Marsh Harrier	*Circus ranivorus*	Southern, central, and eastern Africa from South Africa north to Sudan.
African Pygmy Falcon	*Polihierax semitorquatus*	Eastern and southern Africa.
American Kestrel	*Falco sparverius*	Widely distributed across the Americas; Alaska across northern Canada to Nova Scotia and south throughout North America, into central Mexico, the Baja, and the Caribbean.
Andean Condor	*Vultur gryphus*	Andes Mountains and adjacent Pacific coasts of western South America.

Common Name	Scientific Name	Range
Aplomado Falcon	*Falco femoralis*	Northern Mexico and locally to southern South America.
Bald Eagle	*Haliaeetus leucocephalus*	Most of Canada and Alaska, all the contiguous United States, and northern Mexico.
Barn Owl	*Tyto alba*	Almost anywhere in the world outside polar and desert regions; Asia north of the Alpide belt, most of Indonesia, and the Pacific islands.
Barred Owl	*Strix varia*	Widespread in North America, across most of the eastern half of the continent from Florida to southern Canada.
Bearded Vulture	*Gypaetus barbatus*	Southern Europe, Africa, India, and Tibet.
Black Kite	*Milvus migrans*	Widely distributed across continental Europe, Asia (except the eastern regions), the Near and Middle East, Africa, Indonesia, and Australia.
Black Vulture	*Coragyps atratus*	Southeastern United States to Central Chile and Uruguay in South America. Recently confirmed resident and breeder in New York State.
Black-thighed Falconet	*Microhierax fringillarius*	Malay Peninsula, Java, Sumatra, and Borneo.
Blakiston's Fish-Owl	*Ketupa blakistoni*	Siberia, northeast China, and the Japanese island of Hokkaido.
Boreal Owl/ Tengmalm's Owl	*Aegolius funereus*	Subalpine and boreal forests around the globe.

Common Name	Scientific Name	Range
Broad-winged Hawk	*Buteo platypterus*	Most of eastern North America and west to Alberta and Texas; migrate south to winter in the neotropics from Mexico to Southern Brazil. Many of the subspecies in the Caribbean are endemic and most do not migrate.
Burrowing Owl	*Athene cunicularia*	North America across the grassland regions of southern Alberta, Saskatchewan, and Manitoba. All U.S. states west of the Mississippi Valley; breed south through the western and midwestern states. A separate subspecies is found in Florida and the Caribbean Islands.
California Condor	*Gymnogyps californianus*	The Grand Canyon area, Zion National Park, western coastal mountains of California, and northern Baja California.
Cinereous Vulture	*Aegypius monachus*	Breeds across southern Europe and Asia from Spain to Korea.
Collared Falconet	*Microhierax caerulescens*	Bangladesh, Bhutan, Cambodia, China, Pakistan, India, Laos, Myanmar, Nepal, Thailand, and Vietnam.
Common Buzzard	*Buteo buteo*	Most of Europe and into Asia.
Cooper's Hawk	*Accipiter cooperii*	Widely distributed in North America, from Canada to Mexico.
Crane-Hawk	*Geranospiza caerulescens*	Mexico and widely distributed throughout South America to northern Argentina.

Common Name	Scientific Name	Range
Crested Caracara	*Caracara cheriway*	Southwestern United States and Florida, Central America, and South America.
Crowned Hawk-Eagle	*Stephanoaetus coronatus*	Tropical Africa south of the Sahara.
Egyptian Vulture	*Neophron percopterus*	Southwestern Europe and northern Africa to southern Asia.
Elf Owl	*Micrathene whitneyi*	Southwestern United States and Mexico.
Eurasian Eagle Owl	*Bubo bubo*	Much of Europe and Asia.
Eurasian Hobby	*Falco subbuteo*	Breeds across Europe and Asia; a long-distance migrant, wintering in Africa.
Eurasian Kestrel	*Falco tinnunculus*	Widespread in Europe, Asia, and Africa.
European Sparrowhawk	*Accipiter nisus*	Very widespread throughout most of Europe and Asia.
Golden Eagle	*Aquila chrysaetos*	Once widespread across the Holarctic (northern continents), present in Eurasia, North America, and parts of Africa.
Great Grey Owl	*Strix nebulosa*	Alaska east across Canada, northern Europe, and Asia.
Great Horned Owl	*Bubo virginianus*	Subarctic North America through much of Central America and South America south to Tierra del Fuego.
Greater Spotted Eagle	*Aquila clanga*	Northern Europe across Asia; winters in southeastern Europe, the Middle East, and South Asia.
Greater Yellow-headed Vulture	*Cathartes melambrotus*	Central South America.

Common Name	Scientific Name	Range
Griffon Vulture	Gyps fulvus	Southern Europe, North Africa, and Asia.
Gyrfalcon	Falco rusticolus	Breeds on Arctic coasts and islands of North America, Europe, and Asia. Mainly resident, but some Gyrfalcons disperse more widely after the breeding season or in winter.
Harlan's Hawk	Buteo jamaicensis harlani	Alaska and northwestern Canada; winters on the southern Great Plains.
Harpy Eagle	Harpia harpyja	Rainforests of Central and South America, from southern Mexico south to eastern Bolivia, southern Brazil, and northernmost Argentina.
Harris's Hawk	Parabuteo unicinctus	Southwestern United States through Central America and into drier regions of South America.
Hawaiian Hawk	Buteo solitarius	Endemic to the island of Hawaii; vagrants occasionally wander to Maui, O'ahu, and Kauai.
Himalayan Griffons	Gyps himalayensis	Central Asia to northern India.
Honey Buzzard	Pernis apivorus	Summer migrant to most of Europe and western Asia; winters in tropical Africa.
Indian Vulture	Gyps indicus	India and Southeastern Pakistan.
Indian White-backed Vulture	Gyps bengalensis	Southeastern Iran, Afghanistan, and Pakistan through Nepal and India to south central China, Indochina, and northern Malay Peninsula.

Common Name	Scientific Name	Range
King Vulture	*Sarcoramphus papa*	Southern Mexico, throughout Central and South America to northern Argentina.
Lanner Falcon	*Falco biarmicus*	Breeds in Africa, southeast Europe, and into Asia; mainly resident, but some birds disperse more widely after the breeding season.
Lanyu Scops Owl	*Otus elegans botelensis*	Lanyu Island off the southeastern coast of Taiwan.
Lesser Kestrel	*Falco naumanni*	Breeds from the Mediterranean across southern central Asia to China and Bangladesh; winters in Africa and Pakistan.
Marshal Eagle	*Polemateus bellicosus*	Senegal to Somalia and Cape Province.
Mauritius Kestrel	*Falco punctatus*	Endemic to the forests of Mauritius.
Merlin	*Falco columbarius*	Breeds in northern North America, Europe, and Asia; migratory after the breeding season.
Mexican Spotted Owl	*Strix occidentalis lucida*	Southern Utah and Colorado to the mountains of Arizona, New Mexico, west Texas, and the mountains of northern and central Mexico.
New Zealand Falcon	*Falco novaeseelandiae*	Endemic to New Zealand.
Northern Goshawk	*Accipiter gentilis*	Temperate regions of the northern hemisphere; mainly resident, but birds from colder regions of north Asia and Canada migrate south for the winter.

Common Name	Scientific Name	Range
Northern Harrier	*Circus cyaneus*	Breeds throughout the northern parts of the northern hemisphere in Canada and the northernmost United States, and in northern Eurasia. Migrates south for the winter.
Northern Hawk-Owl	*Surnia ulula*	Boreal forests of North America and Eurasia.
Northern Pygmy Owl	*Glaucidium gnoma*	Foothills and mountains of western North America.
Northern Spotted Owl	*Strix occidentalis caurina*	Pacific coast of North America from extreme southern British Columbia to Marin County in northern California.
Ornate Hawk-Eagle	*Spizaetus ornatus*	Mexico to Columbia and Guyanas to northern Argentina.
Osprey	*Pandion haliaetus*	All continents except Antarctica, and in South America only a nonbreeding migrant.
Palm-nut Vulture	*Gypohierax angolensis*	Breeds in forest and savannah across sub-Saharan Africa.
Peregrine Falcon	*Falco peregrinus*	Nearly everywhere except extreme polar regions, very high mountains, and most tropical rainforests. Breeding range includes land regions from the Arctic tundra to the Tropics.
Red-headed Vulture	*Sarcogyps calvus*	Pakistan to Yunnan, Indochina, and the Malay Peninsula.
Red-footed Falcon	*Falco vespertinus*	Eastern Europe and west, central, and north-central Asia, from Estonia and Hungary to extreme northwestern China; winters in southwest Africa, from Angola, Namibia, and north South Africa through Botswana to Zimbabwe and Zambia.

Common Name	Scientific Name	Range
Red Kite	*Milvus milvus*	Western Palearctic region in Europe and northwest Africa.
Red-tailed Hawk	*Buteo jamaicensis*	Breeds throughout most of North America, from western Alaska and northern Canada to Panama and the West Indies.
Rough-legged Hawk	*Buteo lagopus*	Breeds in northernmost Europe, Asia, and North America; migrates farther south in winter.
Saker Falcon	*Falco cherrug*	Breeds from eastern Europe across Asia to Manchuria. Mainly migratory except in the southernmost parts of its range, wintering in Ethiopia, the Arabian peninsula, northern India, and western China.
Savanna Hawk	*Buteogallus meridionalis*	Panama and Trinidad south to Bolivia, Uruguay, and central Argentina.
Scops Owl	*Otus scops*	Southern Europe, parts of North Africa, Asia Minor east to Central Asia. Many populations migratory, moving to Africa south of the Sahara in winter.
Secretary Bird	*Sagittarius serpentarius*	Widespread in Africa south of the Sahara.
Sharp-shinned Hawk	*Accipiter striatus*	Widespread in North America, Central America, South America, and the Greater Antilles.
Short-eared Owl	*Asio flammeus*	All continents except Antarctica and Australia. Partly migratory, moving south in winter from the northern parts of its range.

Common Name	Scientific Name	Range
Snail Kite	*Rostrhamus sociabilis*	Breeds in tropical South America, the Caribbean, and central and southern Florida in the United States. Resident year-round in most of its range, but the southernmost population migrates north in winter and the Caribbean birds disperse widely outside the breeding season.
Snowy Owl	*Bubo (=Nyctea) scandiacus*	North of the Arctic Circle south through Canada and northernmost Eurasia, with irruptions farther south in some years.
Spanish Imperial Eagle	*Aquila adalberti*	Central and southwest Spain, Portugal, and possibly northern Morocco.
Sparrow Hawk	*Accipiter nisus*	Widespread throughout temperate and subtropical parts of the Old World. Birds from colder regions of northern Europe and Asia migrate south for the winter, some to North Africa (some as far as equatorial east Africa) and India.
Steppe Buzzard	*Buteo buteo vulpinus*	Eastern Europe eastward to the Far East, excluding Japan. A long-distance migrant wintering in Africa, India, and southeastern Asia.
Steppe Eagle	*Aquila nipalensis*	Breeds from Romania east through the south Russian and Central Asian steppes to Mongolia. The European and Central Asian birds winter in Africa, and the eastern birds in India.

Common Name	Scientific Name	Range
Striated Caracara	*Phalcoboenus australis*	Falkland Islands and, rarely, Tierra del Fuego.
Swainson's Hawk	*Buteo swainsoni*	Breeds in prairie and dry grasslands in western North America. A long-distance migrant, wintering in Argentina.
Swamp Harrier	*Circus approximans*	Australasia and the South Pacific; common in New Zealand.
Turkey Vulture	*Cathartes aura*	Throughout the Americas from southern Canada to Cape Horn.
Vermiculated Fishing Owl	*Scotopelia bouvieri*	Angola, Cameroon, Central African Republic, Republic of the Congo, Democratic Republic of the Congo, Gabon, and Nigeria.
White-tailed Sea Eagle	*Haliaeetus albicilla*	Mostly resident in northern Europe and northern Asia.

References

We turned to these two excellent books when we were working on many answers, so they are not listed under specific questions:

Berger, C. 2005. *Owls*. Mechanicsburg, Pa.: Stackpole Books.

Brown L. and D. Amadon. 1968. *Eagles, Hawks and Falcons of the World*. Middlesex, Great Britain: Country Life Books.

Our source for spelling and punctuation of common and scientific names is *The Howard and Moore Complete Checklist of the Birds of the World*, 3rd ed., ed. E. C. Dickinson. Princeton: Princeton University Press, 2003.

Preface

Bodio, S.1994. *A Rage for Falcons*. New York: Schocken Press.

Cade, T. 1982. *The Falcons of the World*. Ithaca, N.Y.: Cornell University Press.

Fox, N. 1995. *Understanding the Bird of Prey*. Blaine, Wash.: Hancock House.

Chapter 1: Raptor Basics

Question 2: Where in the world are raptors found?

Audubon study. 2009. State of the Birds, http://www.stateofthebirds .org/pdf_files/State_of_the_Birds_2009.pdf; accessed April 4, 2009.

Winn, M. 1999. Red Tails in Love: A Wildlife Drama in Central Park. New York: Vintage.

Question 3: What is the connection between "raptor" dinosaurs and modern birds of prey?

A small derived theropod from Öösh, early Cretaceous, Baykhangor Mongolia. American Museum Novitiates, no. 3557, http://hdl .handle.net/2246/5845; accessed May 15, 2009.

Hackett, S. J., et al. 2008. A phylogenomic study of birds reveals their evolutionary history. *Science* 320:1763–1768.

New specimens of Microraptor zhaoianus (Theropoda, Dromaeosauridae) from northeastern China. American Museum Novitiates; no. 3381, http://hdl.handle.net/2246/2870; accessed March 21, 2009.

Novas, F. E., and D. Pol. 2005. New evidence on deinonychosaurian dinosaurs from the Late Cretaceous of Patagonia. *Nature* 433:858–861.

Padian, K., A. J. de Ricqlès, and J. R. Homer. 2001. Dinosaaurian growth rates and bird origins. *Nature* 412:405–408.

Vinther, J., et al. 2008. The colour of fossil feathers. *Biology Letters* 4:522–525.

Question 4: How are raptors classified?

Griffiths, C. S., et al. 2007. Phylogeny, diversity, and classification of the Accipitridae based on DNA sequences of the RAG-1 exon. *Journal of Avian Biology* 38:587–602.

Hackett, S. J., et al. 2008. A phylogenomic study of birds reveals their evolutionary history. *Science* 320:1763–1768.

Mertz, L. A., G. F. Barrowclough, and J. G. Groth. 2004. The phylogeny of owls and the position of *Xenoglaux*. Poster presented at *American Ornithologists' Union meeting, Quebec, Canada*.

Sibley, C. G., and J. E. Ahlquist. 1972. *A Comparative Study of the Egg-White Proteins of Non-Passerine Birds*. New Haven, Conn.: Peabody Museum of Natural History, Yale University.

Sibley, C. G., and B. L. Monroe. 1990. *Distribution and Taxonomy of the Birds of the World*. New Haven, Conn.: Yale University Press.

Question 5: What are the differences between falcons and hawks?

Hackett, S. J., et al. 2008. A phylogenomic study of birds reveals their evolutionary history. *Science* 320:1763–1768.

Question 6: Are all owls nocturnal?

Fishing Owls, Eagle Owls and the Snowy Owl. Fishowls.com, http://www.fishowls.com/Literature/Fogden%201992%20Fishing%20Owls%20(Chapter%20in%20Owls%20of%20the%20World).pdf; accessed March 3, 2009.

Slaght, J. C., and S. G. Surmach. 2008. Biology and conservation of Blakiston's Fish-Owls (*Ketupa blakistoni*) in Russia: A review of the primary literature and an assessment of the secondary literature. *Journal of Raptor Research* 42:29–37.

Question 7: Are eagles the largest raptors?

Conservation status of the eagles of the world. Eagle Conservation Alliance (ECA), http://www.eagleconservationalliance.org/eagles_worldwide.html; accessed March 3, 2009.

Question 11: How long do raptors live in the wild?

Kirschbaum, K. Family Accipitridae. Animal Diversity Web, www.animaldiversity.ummz.edu/site/accounts/information/accipitridae.html; accessed February 14, 2009.

Snyder, H. 2001. Hawks and allies. In *The Sibley Guide to Bird Life and Behavior,* ed. C. Elphick, J. Dunning, and D. Sibley, 212–224. New York: Alfred A. Knopf.

Thiollay, J. 1994. Family Accipitridae (hawks and eagles). In *Handbook of the Birds of the World,* vol. 2, ed. J. del Hoyo, A. Elliot, and J. Sargatal, 52–105. Barcelona: Lynx Editions.

Welty, J. C. 1982. *The Life of Birds.* New York: Harcourt Brace.

Chapter 2: Raptor Bodies

Question 1: How do male and female raptors differ?

Mosher, J. A., and P. F. Matray. 1974. Size dimorphism: a factor in energy savings for broad-winged hawks. *Auk* 91:325–341.

Question 2: What do birds of prey eat?

Olsen, J., et al. 2008. Dietary shifts based upon prey availability in Peregrine Falcons and Australian Hobbies breeding near Canberra, Australia. *Journal of Raptor Research* 42:125–137.

Palm-nut Vulture. Kenya Birds, http://www.kenyabirds.org.uk/vult-pn.htm; accessed April 25, 2009.

Seaton, E., et al. 2008. Breeding season diet and prey selection of the New Zealand Falcon (*Falco novaeseelandiae*) in a plantation forest. *Journal of Raptor Research* 42:256–264.

Thomson, A. L., and R. E. Moreau. 1957. Feeding habits of the Palm-nut Vulture *Gypoheerax. Ibis* 99:608–613.

Question 3: How much does a raptor eat in a day?

Overskaug, K., P. Sunde, and E. Kristiansen. 1997. Subcutaneous fat accumulation in Norwegian owls and raptors. *Ornis Fennica* 74:29–37.

Tieleman, I., et al. 2009. Genetic modulation of energy metabolism in birds through mitochondrial function. *Proceedings of the Royal Society B: Biological Sciences* 276:1685–1693.

Question 4: How do raptors digest their food?

Barton, N.W.H. 1993. The influence of gut morphology on digestion time in raptors. *Comparative Biochemistry and Physiology A: Comparative Physiology* 105:571–578.

Joseph, V. 2006. Raptor medicine: An approach to wild, falconry, and educational birds of prey. *Veterinary Clinics of North America: Exotic Animal Practice* 9:321–345.

Kirkwood, J. K. 1979. The partition of food energy for existence in the Kestrel *Falco tinnunculus* and the Barn Owl *Tyto alba*. *Comparative Biochemistry and Physiology A: Physiology* 63A:495–498.

Tollan, A. M. 1988. Maintenance energy requirements and energy assimilation efficiency of the Australasian Harrier. *Ardea* 76:181–186.

Question 5: Can the species of a raptor be identified by its waste pellets?

Trapani, J., et al. 2006. Precision and consistency of the taphonomic signature of predation by Crowned Hawk-Eagles (*Stephanoaetus coronatus*) in Kibale National Park, Uganda. *Palaios* 21:114–131.

Question 6: Are raptors warm- or cold-blooded?

Arad, Z., and M. H. Bernstein. 1988. Temperature regulation in Turkey Vultures. *Condor* 90:913–919.

Arad, Z., U. Midtgard, and M. H. Bernstein. 1989. Thermoregulation in Turkey Vultures: Vascular anatomy, arteriovenous head exchange, and behavior. *Condor* 91:505–514.

Bartholomew, G. A., and T. J. Cade. 1957. The body temperature of the American Kestrel, *Falco sparverius*. *Wilson Bulletin* 69:149–154.

Mosher, J. A., and C. M. White. 1978. Falcon temperature regulation. *Auk* 95:80–84.

Ward, J., et al. 2008. Why do vultures have bald heads? The role of postural adjustment and bare skin areas in thermoregulation. *Journal of Thermal Biology* 33:168–173.

Witter, M. S., and I. C. Cuthill. 1993. The ecological costs of avian fat storage. *Proceedings of the Royal Society B: Biological Sciences* 340:73–92.

Question 8: Do raptors molt?

Bald eagle. Restoration Notebook, http://www.evostc.state.ak.us/ Universal/Documents/Publications/RestorationNotebook/RN _baldeagle.pdf; accessed August 23, 2009.

Cannell, P. 1984. A revised age/sex key for Mourning Doves, with comments on the definition of molt. *Journal of Field Ornithology* 55:112–114.

Question 11: How well do birds of prey hear?

Brooke, M., and T. Birkhead. 1991. *Cambridge Encyclopedia of Ornithology*. London: Cambridge University Press.

Fox, N. 1995. *Understanding the Bird of Prey*. Blaine, Wash.: Hancock House.

Gutfreund, Y., and E. I. Knudsen. 2006. Adaptation in the auditory space map of the barn owl. *Neurophysiology* 96:813–825.

Rice, W. R. 1982. Acoustical location of prey by the Marsh Hawk: Adaptation to concealed prey. *Auk* 99:403–413.

Walsh, S. A., et al. 2009. Inner ear anatomy is a proxy for deducing auditory capability and behavior in reptiles and birds. *Proceedings of the Royal Society B: Biological Sciences* 276:1355–1360.

Witten, I. B., and E. I. Knudsen. 2005. Why seeing is believing: Merging auditory and visual worlds. *Neuron* 48:489–496.

Question 12: How well can raptors see?

Andison, M. E., J. G. Sivak, and D. M. Bird. 1992. The refractive development of the eye of the American Kestrel (*Falco sparverius*): A new avian model. *Journal of Comparative Physiology A: Neuroethology, Sensory, Neural, and Behavioral Physiology* 170:565–574.

Bach, M., et al. 2000. Standard for pattern electroretinography. *Documenta Ophthalmologica* 101:11–18.

Duke-Elder, S. 1958. *The Eye in Evolution*. London: Henry Kimpton.

Fite, K. V., and S. Rosenfield-Wessels. 1975. A comparative study of deep avian foveas. *Brain, Behavior and Evolution* 12:97–115.

Fox, N. 1995. *Understanding the Bird of Prey*. Blaine, Wash.: Hancock House.

Gaffney, M., and W. Hodos. 2003. The visual acuity and refractive state of the American kestrel (*Falco sparverius*). *Vision Research* 43:2053–2059.

Hodos, W. 1993. The visual capabilities of birds. In *Vision, Brain, and Behavior in Birds*, ed. H. P. Zeigler and H. J. Bischof, 63–76. Cambridge, Mass.: MIT Press.

Inzunza, O., et al. 2005. Topography and morphology of retinal ganglion cells in Falconiforms: A study on predatory and carrion-eating birds. *Anatomical Record* 229:271–277.

Jones, M. P., K. E. Pierce, and D. Ward. 2007. Avian vision: A review of form and function with special consideration to birds of prey. *Journal of Exotic Pet Medicine* 16:69–87.

Korpimaki, E., V. Koivunen, and H. Hakkarainen. 1996. Microhabitat use and behavior of voles under weasel and raptor predation risk: Predator facilitation? *Ecology* 7:30–34.

Murphy, C. J. 1987. Raptor ophthalmology. Compendium on Continuing Education for the Practicing Veterinarian 9:241–260.

Murphy, C. J., M. Howland, and H. C. Howland. 1995. Raptors lack lower-field myopia. *Vision Research* 35:1153–1155.

The nervous system and senses. Ornithology.com, http://www.ornithology.com/lectures/Senses.html; accessed December 18, 2008.

Pauli, A., et al. 2007. Clinical techniques: considerations for release of raptors with ocular disease. *Journal of Exotic Pet Medicine* 16:101–103.

Schwab, L. R., and D. Maggs. 2004. The falcon's stoop. *British Journal of Ophthalmology* 88:4.

Tucker, V. A. 2000. The deep fovea, sideways vision, and spiral flight paths in raptors. *Journal of Experimental Biology* 203:3745–3754.

The Ultimate Guide: Birds of Prey: Golden Eagle Vision. Video. Howstuffworks.com, http://videos.howstuffworks.com/discovery/30495-the-ultimate-guide-birds-of-prey-golden-eagle-vision-video.htm.

Walls, G. L. 1942. *The Vertebrate Eye and Its Adaptive Radiation.* Bloomfield, Mich.: Cranbrook Institute of Science.

Wiebe, K. J., and A. Basu. 1997. Modelling ecologically specialized biological visual systems. *Pattern Recognition* 30:1687–1703.

Question 13: Do raptors have a keen sense of smell?

Fox, Nick. 1995. *Understanding the Bird of Prey.* Canada: Hancock House.

Houston, D. C. 1986. Scavenging efficiency of turkey vultures in tropical forest. *Condor* 88:318–323.

McShea, W. J., et al. 2000. An experiment on the ability of free-ranging turkey vultures (*Cathartes aura*) to locate carrion by chemical cues. *Chemoecology* 10:49–50.

Snyder, N., and H. Snyder. 1991. *Raptors: North American Birds of Prey.* Stillwater, Minn.: Voyageur Press.

Stager, K. E. 1964. The role of olfaction in food location by the Turkey Vulture (*Cathartes aura*). *Los Angeles County Museum Contributions to Science* 81:1–63.

Chapter 3

Question 1: How intelligent are birds of prey?

Bednarz, J. C. 1988. Cooperative hunting Harris' Hawks (*Parabuteo unicinctus*). *Science* 239:1525–1527.

Garamszegi, L. Z., A. P. Moller, and J. Erritzoe. 2002. Coevolving avian eye size and brain size in relation to prey capture and nocturnality. *Proceedings of the Royal Society B: Biological Sciences* 269:961–967.

Question 3: How fast can a falcon dive?

Cade, T. 1982. *The Falcons of the World.* New York: Cornell University Press.

Clark, C. J. 2009. Courtship dives of Anna's hummingbird offer insights into flight performance limits. *Proceedings of the Royal Society B: Biological Sciences* 276:3047–3052.

Franklin, Ken. 1999. Vertical flight. *NAFA Journal* 38:68–72.

Raptor adaptations. Delaware Valley Raptor Center, http://www.dvrconline.org/raptoradapt.html; accessed April 25, 2009.

Question 4: How far can a raptor fly?

Bildstein, K. L. 2006. *Migrating Raptors of the World.* New York: Cornell University Press.

Cade, T. J., and L. Greenwald. 1966. Nasal salt secretion in Falconiform birds. *Condor* 68:338–350.

Gorney, E., and Y. Yom-Tov. 2008. Fat, hydration condition, and molt of Steppe Buzzards *Buteo buteo vulpinus* on spring migration. *Ibis* 136:185–192.

England, A. S., M. C. Bechard, and C. S. Houston. 1997. Swainson's Hawk (*Buteo swainsoni*). No. 265 in *The Birds of North America,* ed. A. Poole and F. Gill. Philadelphia: Academy of Natural Sciences.

Peregrine biology: Habitat and distribution. Canadian Peregrine Foundation, http://www.peregrine-foundation.ca/info/habitat.html; accessed March 24, 2009.

Shoemaker, V. H. 1972. Osmoregulation and excretion in birds. In *Avian Biology,* vol. 2, ed. D. S. Farmer and J. R. King, 527–574. New York: Academic Press.

Sparr, R. 1995. Flight behavior of Steppe Buzzards (*Buteo buteo vulpinus*) during spring migration in southern Israel: A tracking-radar study. *Israel Journal of Zoology* 41:489–500.

Strandberg, R., et al. 2009. Converging migration routes of Eurasian obbies *Falco subbuteo* crossing the African equatorial rain forest. *Proceedings of the Royal Society B: Biological Sciences* 276:727–733.

White, C., et al. Peregrine falcon. The Birds of North America Online, http://bna.birds.cornell.edu/bna/species/660/articles/introduction; accessed March 17, 2009.

Zalles, J. I., and K. L. Bildstein, eds. 2000. *Raptor Watch: A Global Directory of Raptor Migration Sites.* Vol. 9, BirdLife Conservation Series. Washington, D.C.: Smithsonian Institution Press.

Question 5: Do all raptors migrate?

Both, C. et al. 2009. Avian population consequences of climate change are most severe for long-distance migrants in seasonal habitats. *Proceedings of the Royal Society B: Biological Sciences* 277: 1259–1266.

Newton, I. 2008. *The Migration Ecology of Birds.* London, U.K.: Academic Press.

Strandberg, R., et al. 2009. How hazardous is the Sahara Desert crossing for migratory birds? Indications from satellite tracking of raptors. *Proceedings of the Royal Society B: Biological Sciences.* Published online before print, December 2, 2009.

Zalles, J. I., and K. L. Bildstein, eds. 2000. *Raptor Watch: A Global Directory of Raptor Migration Sites.* Vol. 9, BirdLife Conservation Series. Washington, D.C.: Smithsonian Institution Press.

Question 6: How do raptors find their way during migration?

Bildstein, K. L. 2006. *Migrating Raptors of the World: Their Ecology and Conservation.* New York: Cornell University Press.

Davies, H., and C. A. Butler. 2008. *Do Butterflies Bite?* Piscataway, N.J.: Rutgers University Press.

Emlen, S. 1967. Migratory orientation in the Indigo Bunting, *Passerina cyanea. Auk* 84:309–342.

Gorney, E., and Y. Yom-Tov. 2008. Fat, hydration condition, and molt of Steppe Buzzards *Buteo buteo vulpinus* on spring migration. *Ibis* 136:185–192.

Kerlinger, P. 1989. *Flight Strategies of Migrating Hawks.* Chicago: University of Chicago Press.

Wiltschko, W., et al. 2009. Avian orientation: the pulse effect is mediated by the magnetite receptors in the upper beak. *Proceedings of the Royal Society B: Biological Sciences* 276:2227–2232.

Yosef, R., P. Tryjanowski, and K. L. Bildstein. 2002. Spring migration of adult and immature buzzards (*Buteo buteo*) through Elat, Israel: timing and body size. *Journal of Raptor Research* 36:115–120.

Question 8: Are birds of prey social or loners?

Craighead, J. J., and F. C. Craighead Jr. 1969. *Hawks, Owls and Wildlife.* New York: Dover Publications.

Question 10: Are raptors always aggressive?

Grier, J. M. 1969. Bald Eagle behaviour and productivity responses to climbing to nests. *Journal of Wildlife Management* 33:961–966.

Knight, R. L. 1984. Responses of nesting Ravens to human being in areas of different human densities. *Condor* 86:345–346.

Sherrod, S. K., E. M. White, and E.S.L. Williamson. 1977. Biology of the Bald Eagle (*Haliaetus leucocephalus alascanus*) on Amichitka Island, Alaska. *Living Bird* 15:143–182.

Chapter 4: Raptor Reproduction

Question 2: At what time of year do birds of prey mate?

Reudink, M. W., et al. 2009. Non-breeding season events influence sexual selection in a long-distance migratory bird. *Proceedings of the Royal Society B: Biological Sciences* 276:1619–1626.

Question 3: How do raptors mate?

Brennan, P.L.R., et al. 2007. Coevolution of male and female genital morphology in waterfowl. *PLoS ONE* 2:e418.

Briskie, J. V., and R. Montgomerie. 1997. Sexual selection and the intromittent organ of birds. *Journal of Avian Biology* 28:73–86.

Calhim, S., and T. R. Birkhead. 2009. Intraspecific variation in testis asymmetry in birds: evidence for naturally occurring compensation. *Proceedings of the Royal Society B: Biological Sciences* 276:2279–2284.

Gowaty, P. A., and N. Buschhaus. 1998. Ultimate causation of aggressive and forced copulation in birds: Female resistance, the CODE hypothesis, and social monogamy. *American Zoologist* 38:207–225.

Lombardo, M. P., P. A. Thorpe, and H. W. Power. 1999. The beneficial sexually transmitted microbe hypothesis of avian copulation. *Behavioral Ecology* 10:333–337.

Ovarian dynamics and follicle development in Aves. Avianova, http://www.nd.edu/~avianova; accessed March 12, 2009.

Peterson, A. D., M. P. Lombardo, and H. W. Power. 2001. Left-sided directional bias of cloacal contacts during tree swallow copulations. *Animal Behavior* 62:739–741.

Question 4: Are raptors monogamous?

Arroyo, B. E. 1999. Copulatory behavior of semi-colonial Montagu's Harriers. *Condor* 101:340–346.

Arsenault, D. P., P. B. Stacey, and G. A. Hoelzer. 2002. No extra-pair fertilization in Flammulated Owls despite aggregated nesting. *Condor* 104:197–201.

Balgooyen, T. G. 1976. Behavior and ecology of the American Kestrel *Falco sparverius*. *Auk* 104:321–324.

Birkhead, T. R., and C. M. Lessels. 1988. Copulation behavior of the osprey *Pandion haliaetus*. *Animal Behavior* 36:1672–1682.

Birkhead, T. R., and A. P. Moller. 1992. *Sperm Competition in Birds*. London: Academic Press.

Bray, O. E., J. J. Kennelly, and J. L. Guarino. 1975. Fertility of eggs produced on territories of vasectomized Red-winged Blackbirds. *Wilson Bulletin* 87:187–196.

Eldegard, K., and G. A. Sonerud. 2009. Female offspring desertion and male-only care increase with natural and experimental increase in food abundance. *Proceedings of the Royal Society B: Biological Sciences* 276:1713–1721.

Hsu, Y., et al. 2005. High frequency of extrapair copulation with low level of extrapair fertilization in the Lanyu Scops Owl *Elegans botelensis*. *Journal of Avian Biology* 37:36–40.

Korpimaki, E., et al. 1996. Copulatory behaviour and paternity determined by DNA fingerprinting in kestrels: Effects of cyclic food abundance. *Animal Behavior* 51:945–955.

Möller, A. P. 1987. Copulation behavior in the goshawk, *Accipiter gentilis*. *Animal Behavior* 35:755–763.

Mougeot, F. 2004. Breeding density, cuckoldry risk and copulation behavior during the fertile period in raptors: a comparative analysis. *Animal Behavior* 67:1067–1076.

Mougeot, F., B. E. Arroyo, and V. Bretagnolle. 2001. Decoy presentations as a means to manipulate the risk of extrapair copulation: An experimental study in a semicolonial raptor, the Montagu's Harrier (*Circus pygargus*). *Behavioral Ecology* 12:1–7.

Negro, J. J., and J. M. Grande. 2001. Territorial signaling: A new hypothesis to explain frequent copulation in raptorial birds. *Animal Behavior* 62:803–809.

Simmons, R. E. 1990. Copulation patterns of African marsh harriers: Evaluating the paternity assurance hypothesis. *Animal Behavior* 40:1151–1157.

———. 2000. *Harriers of the World*. London: Oxford University Press.

Widén, P., and M. Richardson. 2000. Copulation behavior in the osprey in relation to breeding density. *Condor* 102:349–354.

Question 5: Do raptors of one species mate with other species?

Eberhard, W. G. 1985. *Sexual Selection and Animal Genitalia.* Cambridge, Mass.: Harvard University Press.

Fox, N. 1995. *Understanding the Bird of Prey.* Blaine, Wash.: Hancock House.

Hybrid falcons. Raptors, http://www.pauldfrost.co.uk/hybridf.html; accessed November 2, 2008.

Nittinger, F., et al. 2007. Phyleogeography and population structure of the saker falcon (*Falco cherrug*) and the influence of hybridization: Mitochondrial and microsatellite data. *Molecular Ecology* 16:1497–1517.

Oliphant, L. W. 1991. Hybridization between a Peregrine Falcon and a Prairie Falcon in the wild. *Journal of Raptor Research* 25:36–39.

Question 6: How is artificial insemination practiced with raptors?

Gill, P. Modern captive breeding. The Falconers' Web, http://www.falconers.com/articles/captive_breeding_1; accessed March 21, 2009.

Question 7: Do all birds of prey make nests?

Kristan, D. M., R. T. Golightly Jr., and S. M. Tomkiewicz Jr. 1996. A solar-powered transmitting video camera for monitoring raptor nests. *Wildlife Society Bulletin* 24:284–290.

Peregrine biology: habitat and distribution. Canadian Peregrine Foundation, http://www.peregrine-foundation.ca/info/habitat.html; accessed March 24, 2009.

Sattler, H. R. 1995. *The Book of North American Owls.* New York: Clarion Books.

White, C., et al. 2002. Peregrine Falcon (Falco peregrinus). The Birds of North America, http://bna.birds.cornell.edu/BNA/account/Peregrine_Falcon/; accessed March 18, 2009.

Woodford, J. E., C. A. Eloranta, and A. Rinaldi. 2008. Nest density, productivity, and habitat selection of Red-shouldered Hawks in a contiguous forest. *Journal of Raptor Research* 42:79–86.

Yi-Qun, Wu, et al. 2008. Breeding biology and diet of the Long-legged Buzzard (Buteo rufinus) in the Eastern Junggar Basin of northwestern China. *Journal of Raptor Research* 42:273–280.

Question 9: How many eggs do various species of raptors lay?

Weidensaul, S. 2007. *Of a Feather.* New York: Harcourt.

Question 10: What do raptor eggs look like?

Weidensaul, S. 2007. *Of a Feather.* New York: Harcourt.

Chapter 5: Dangers and Defenses

Question 2: How do prey animals defend themselves against raptors?

Dugatkin, L. A., and J-G. J. Godin. 1992. Prey approaching predators: A cost-benefit perspective. *Annals of Zoology Fennici* 29:233–252.

Griesser, M. 2009. Mobbing calls signal predator category in a kin group-living bird species. *Proceedings of the Royal Society B: Biological Sciences* 276:2887–2892.

Hingee, M., and R. D. Magrath. 2009. Flights of fear: A mechanical wing whistle sounds the alarm in a flocking bird. *Proceedings of the Royal Society B: Biological Sciences* 276:4173–4179.

Ito, R., and A. Mori. 2009. Vigilance against predators induced by eavesdropping on heterospecific alarm calls in a non-vocal lizard *Oplurus cuvieri cuvieri* (Reptilia: Iguania). *Proceedings of the Royal Society B: Biological Sciences* 277:1275–1280.

Yorzinski, J. L. and G. L. Patricelli. 2009. Birds adjust acoustic directionality to beam their anti-predator calls to predators and conspecifics. *Proceedings of the Royal Society B: Biological Sciences* 277: 923–932.

Question 3: How do raptors defend themselves?

Boal, C. W. 2001. Agonistic behavior of Cooper's Hawks. *Journal of Raptor Research* 35:253–256.

Question 4: What illnesses occur in wild raptors?

Pain, D. J., et al. 2008. The race to prevent the extinction of South Asian vultures. *Bird Conservation International* 18:S30–S48.

Shannon, L. M., et al. 1988. Serological survey for rabies antibodies in raptors from California. *Journal of Wildlife Diseases* 24:264–267.

What is a raptor rehabilitator and why? Wingmasters.net, www .wingmasters.net/raptor.htm; accessed February 27, 2009.

Wildlife orphans? Think before you act. Ohio Department of Natural Resources, http://www.dnr.state.oh.us/tabid/5665/Default.aspx #rabies; accessed February 27, 2009.

Question 5: What injuries are common among wild raptors?

Castle, J. H. 2007. Nocturnal behavior of raptors at wind energy facilities. Master's thesis, Sonoma State University.

Dwyer, J. F., and R. W. Mannan. 2007. Preventing raptor electrocutions in an urban environment. *Journal of Raptor Research* 41:259–267.

Pauli, A., et al. 2007. Clinical techniques: considerations for release of raptors with ocular diseases. *Journal of Exotic Pet Medicine* 16:101–103.

Redig, P. T., and G. E. Duke. The effect and value of raptor rehabilitation in North America. University of Minnesota, http://www.cvm.umn.edu/raptor/about/publications/raptorrehabilitation/home.html; accessed March 23, 2009.

Results of the 2001 Raptor Electrocution Reduction Program. Hawk Watch International, http://hawkwatch.org/rw_issue.php?id=51; accessed March 23, 2009.

Schmidt-French, B., and C. A. Butler. 2009. *Do Bats Drink Blood?* Piscataway, N.J.: Rutgers University Press.

Sinclair, K. 2001. *Status of Avian Research at the National Renewable Energy Laboratory.* Golden, Colo.: National Renewable Energy Laboratory.

Smallwood, K. S., and C. G. Thelander. 2004. *Developing Methods to Reduce Bird Mortality in the Altamont Pass Wind Resource Area.* Sacramento, Calif.: California Energy Commission.

Zapped. Audubon Magazine, http://audubonmagazine.org/incite/incite0001.html; accessed April 29, 2009.

Question 6: What other dangers do raptors face as a result of development and population growth?

2008 Final recovery plan for the Northern Spotted Owl. U.S. Fish and Wildlife Service, http://www.fws.gov/Pacific/ecoservices/endangered/recovery/NSORecoveryPlanning.htm; accessed March 11, 2009.

Birds: California Condor. San Diego Zoo, http://www.sandiegozoo.org/animalbytes/t-condor.html; accessed April 18, 2009.

Clark, H. O., and D. L. Plumpton. 2005. A simple one-way door design for passive relocation of Western Burrowing Owls. *California Fish and Game* 91:286–289.

Trulio, L. A. (1995): Passive relocation: A method to preserve burrowing owls on disturbed sites. *Journal of Field Ornithology* 66:99–106.

Question 7: Does lead in the environment affect raptors?

Get the lead out. Minnesota Pollution Control Agency, http://www.pca.state.mn.us/oea/reduce/sinkers.cfm; accessed February 28, 2009.

Question 8: Has DDT affected birds of prey?

Florida Fish and Wildlife Conservation Commission Osprey nest removal policies. Ospreys.com, www.ospreys.com/Osprey_policies.pdf; accessed February 24, 2009.

Peregrine Falcons aka Duck Hawks, Bullet Hawks. AvianWeb.com, http://www.avianweb.com/peregrinefalcons.html; accessed March 11, 2009.

The 2006 Osprey Project in New Jersey. State of New Jersey, http:// www.nj.gov/; accessed February 24, 2009.

Question 9: Which raptors are particularly vulnerable to environmental toxins?

BirdLife International. 2004. Birds in Europe: Population Estimates, Trends, and Conservation Status. No. 12, BirdLife Conservation Series. Cambridge: BirdLife International.

Blanco, G., et al. 2007. Geographical variation in cloacal microflora and bacterial antibiotic resistance in a threatened avian scavenger in relation to diet and livestock farming practices. *Environmental Microbiology* 9:1738–1749.

Choresh, Y. *Protecting the Griffon Vulture population in Israel, 2006–2009.* http://bioteach.snunit.k12.il/upload/.mail/nesher.ppt#30; accessed August 28, 2009.

Handrinos, G. 1985. The status of vultures in Greece. In *Conservation Studies in Raptors.* No. 5, ICBP Technical Publication Series, ed. I. Newton and R. D. Chancellor, 103–115. Princeton: Princeton University Press.

Lemus, J. A., and G. Blanco. 2009. Cellular and humoral immunodepression in vultures feeding upon medicated livestock carrion. *Proceedings of the Royal Society B: Biological Sciences* 276:2307–2313.

Levy-Yamamori, R. 2003. *Faradis.* Binyamina, Israel: Har-Ya'ar Books.

———. 2005. *Vultures in Red Skies.* Binyamina, Israel: Har-Ya'ar Books.

———. 2008. *Entas.* Binyamina, Israel: Har-Ya'ar Books.

Naoroji, R. 2008. *Birds of Prey of the Indian subcontinent.* London: Christopher Helm.

Poison takes toll on Africa's lions. CBS News, http://www.cbsnews.com/stories/2009/03/26/60minutes/main4894945.shtml; accessed March 31, 2009.

Schmidt-French, B., and C. A. Butler. 2009. *Do Bats Drink Blood?* New Brunswick, N.J.: Rutgers University Press.

Xirouchakis, S. M., and M. Mylonas. 2005. Status and structure of the Griffon Vulture (*Gyps fulvus*) population in Crete. *European Journal of Wildlife Research* 51:223–231.

Yosef, R., and O. Bahat. 2000. Habitat loss and vultures: A case study from Israel. In *Raptors at Risk,* ed. R. D. Chancellor and B. U. Meyburg, 207–212. Blaine, Wash.: Hancock House.

Question 10: Do other environmental toxins endanger birds of prey?

DeSorbo, C. R., and D. C. Evers. 2006. Evaluating exposure of Maine's Bald Eagle population to mercury: Assessing impacts on productivity and spatial exposure patterns. Report BRI 2006–02. Gorham, Maine: BioDiversity Research Institute.

Di Silvestro, R. 1996. Poison in the Pampas: What's killing the Swainson's hawk? *International Wildlife* 26:38–43.

Evers, D. C., and C. DeSorbo. 2008. Assessing mercury exposure and spatial patterns in adult and nestling Bald Eagles in New York State, with an emphasis on the Catskill region. BioDiversity Research Institute, http://www.briloon.org/pub/doc/2008%20BAEA_%204-07–08%20TNC_2008_6.pdf; accessed March 23, 2009.

Evers, D. C., Young-Ji Han, Charles T. Driscoll, Neil C. Kamman, M. Wing Goodale, Kathleen Fallon Lambert, Thomas M. Holsen, Celia Y. Chen, Thomas A. Clair, and Thomas Butler. 2007. Biological mercury hotspots in the northeastern United States and southeastern Canada. *BioScience* 57:29–43.

Holt, R. D., and M. Barfield. 2009. Trophic interactions and range limits: The diverse roles of predation. *Proceedings of the Royal Society B: Biological Sciences* 276:1435–1442.

Hooper, M. J., et al. 1999. Monocrotophos and the Swainson's Hawk. *Pesticide Outlook* 10:97–102.

Levy, S. 1997. A hawk's-eye perspective: To Brian Woodbridge, saving rare hawks is a family affair. *Animals,* November–December.

Woodbridge, B., K. K. Finley, and S. T. Seager. 1995. An investigation of the Swainson's hawk in Argentina. *Journal of Raptor Research* 29:202–204.

Chapter 6: Raptor Husbandry

Question 1: What is meant by "husbandry"?

The history of falconry. Virginia Falconers' Association, http://vafalconry.swva.net/History%20of%20Falconry.htm; accessed April 4, 2009.

Lascelles, G. 1971. *The Art of Falconry.* Reprint of Frederick II of Hohenstaufen, *De Arte Venandi cum Avibus.* London: C. W. Daniel.

Question 2: How do zoos and rehabilitation facilities house raptors?

Pay for a cage. Wildcare Inc., http://www.wildcareinc.org/cage.html; accessed April 4, 2009.

Question 3: What does a rehabilitator do with a sick or injured raptor?

Raptor rehabilitation. California Foundation for Birds of Prey, http://
www.cafbp.com; accessed February 23, 2009.
What's a raptor rehabilitator? And why? Wingmasters, http://
wingmasters.net/raptor.htm; accessed February 23, 2009.

Question 4: What is "imprinting"?

Lorenz, K. 2002. *King Solomon's Ring: New Light on Animal Ways.* Florence, Ken.: Routledge.
What's a raptor rehabilitator? And why? Wingmasters, http://www
.wingmasters.net/raptor.htm; accessed September 9, 2009.

Question 7: How long do raptors live in captivity?

The effect and value of raptor rehabilitation in North America.
University of Minnesota Raptor Center, http://www.cvm.umn.
edu/raptor/about/publications/raptorrehabilitation/home.html;
accessed April 4, 2009.
Kirschbaum, K. Family Accipitridae. Animal Diversity Web, www
.animaldiversity.ummz.edu/site/accounts/information/accipitridae
.html; accessed February 14, 2009.
Learn. Carolina Raptor Center, www.carolinaraptorcenter.org/
qa.php; accessed April 4, 2009.
Newton, Ian. 1979. *Population Ecology of Raptors.* Vermillion, S.D.:
Buteo Books.
Pauli, A., et al, 2007. Clinical techniques: considerations for release
of raptors with ocular disease. *Journal of Exotic Pet Medicine,*
16:101–103.
Thiollay, J. 1994. Family Accipitridae (hawks and eagles). In *Handbook
of the Birds of the World,* vol. 2, ed. J. del Hoyo, A. Elliot, and J. Sargatal. 52–105. Barcelona: Lynx, Editions.
Welty, J. C. 1982. *The Life of Birds.* 3rd ed. Philadelphia: Saunders College Publishing.

Chapter 7: Taming and Training
Question 1: What is falconry?

Bodio, S.1994. *A Rage for Falcons.* New York: Schocken Press.
Cade, T., and Digby, R. D. 1982. *The Falcons of the World.* Comstock,
N.Y.: Cornell University Press.
Hunting in Tudor England. Tudorplace.com, http://www.tudorplace
.com.ar/Documents/hunting.htm; accessed March 2, 2009.

Question 2: Where and when did humans begin using captive raptors for hunting?

Falconry, Renaissance style. renaissancefestival.com, http://www .renaissancefestival.com/index.php?option=com_content&task= view&id=13&Itemid=2; accessed September 2, 2009.

Question 3: What role do dogs play in falconry?

Argue, D. Dogs in falconry. The Falconers Web, http://www.falconers .com/articles/dogs3; accessed March 30, 2009.

Falconry/Hawking. Poodle History Project, http://www.poodlehistory .org/PFANDH.HTM; accessed March 30, 2009.

Lascelles, G. 1971. *The Art of Falconry*. Reprint of Frederich II of Hohenstaufen, *De Arte Venandi cum Avibus*. London: C. W. Daniel.

Question 5: Do you always need a license to possess a raptor?

Bird of prey experience day. Eagle Heights, http://www.eagleheights .co.uk/courses.htm; accessed March 30, 2009.

Falconry legislation—Europe. Falconry Pro, http://www.falconrypro .com/legistration-europe.html; accessed August 29, 2009.

International Association for Falconry and Conservation of Birds of Prey. IAF, http://www.i-a-f.org/history.html; accessed August 29, 2009.

List a raptor for sale. Raptors etc., http://www.raptorsforsale.com/ falconry-dogs-for-sale.asp; accessed March 30, 2009.

Chapter 8: Raptors and People

Question 1: Have attitudes about raptors changed over time?

Murphy, R. C. 1950. Frank Michler Chapman, 1864–1945. *Auk* 67:307–317.

Weidensaul, S. 2007. *Birds of a Feather*. New York: Harcourt.

Question 3: Where can I see raptors?

Heintzelman, D. S. 2004. *Guide to Hawk Watching in North America*. Helena, Mont.: Falcon Press.

National Raptor Migration Corridor Project. Kittatinny-Shawangunk, http://www.raptorcorridor.org; accessed April 30, 2009.

Question 5: What attracts raptors to live in cities?

New York bird watching. Birding.com, http://www.birding.com/ wheretobird/newyork.asp; accessed March 4, 2009.

North American rare bird alert telephone listing. American Birding Association, http://www.aba.org/resources/rbaphonelist.html; accessed March 22, 2009.

Peregrine biology: Habitat and distribution. Canadian Peregrine Foundation, http://www.peregrine-foundation.ca/info/habitat. html; accessed March 24, 2009.

Chapter 9: Research and Conservation

Question 1: Why do we need to study raptors?

Banks, P. B., et al. 2004. Dynamic impacts of feral mink predation on vole metapopulations in the outer archipelago of the Baltic Sea. *Oikos* 105:79–88.

Blakiston's Fish Owl Project. http://www.fishowls.com; accessed April 12, 2009.

Craighead, J. J., and F. C. Craighead. 1969. *Hawks, Owls, and Wildlife.* New York: Dover Publications.

Gaston, K. J. 2009. Geographic range limits of species. *Proceedings of the Royal Society B: Biological Sciences* 276:1391–1393.

Gilliam, J. F., and D. F. Fraser. 2001. Movement in corridors: Enhancement by predation, threat, disturbance, and habitat structure. *Ecology* 82:258–273.

Holt, R. D., and M. Barfield. 2009. Trophic interactions and range limits: The diverse roles of predation. *Proceedings of the Royal Society B: Biological Sciences,* 276:1435–1442.

The state of the birds, United States of America, 2009. *The State of the Birds,* http://www.stateofthebirds.org/pdf_files/State_of_the_Birds_2009.pdf; accessed March 21, 2009.

Question 4: How are raptor skins prepared for study or exhibit?

Galeotti, P., et al. 2009. Global changes and animal phenotypic responses: Melanin-based plumage redness of Scops Owls increased with temperature and rainfall during the last century. *Biology Letters* 5:532–534.

Winker, K. 2000. Obtaining, preserving, and preparing bird specimens. *Journal of Field Ornithology* 71:250–297.

Question 5: Are any raptors endangered?

National Eagle repository. U.S. Fish & Wildlife Service, http://www.fws.gov/mountain-prairie/law/eagle; accessed April 25, 2009.

Sidebar: Taxonomy

Bromham, L. 2009. Darwin would have loved DNA. *Biology Letters* 5:503–505.

Sidebar: Bird Strikes

Dolbeer, R. A., S. E. Wright, and E. C. Cleary. 2006. Real birds versus whirly birds: Bird strikes to civil helicopters in the USA, 1990–2005. *Abstracts of the Proceedings of the Eighth Bird Strike Committee USA/ Canada Annual Meeting.* St. Louis, Missouri.

Sidebar: Women in Falconry

A treatyse of fysshynge with an Angle. Renascence Editions, http://darkwing.uoregon.edu/~rbear/berners/berners.html; accessed May 13, 2009.
Ford, E. 1999. *Fledgling Days: Memoir of a Falconer.* Woodstock, N.Y.: Overlook Press.
Hammerstrom, F. 1988. *An Eagle to the Sky.* Essex, Conn.: Lyons Press.
Hunting (falconer). Medieval Women, http://mw.mcmaster.ca/ scriptorium/hunting.html; accessed May 12, 2009.
Macdonald, H. 2006. *Falcon.* London: Reaktion Books.
Otsuka, N. 2006. Falconry: Tradition and acculturation. *International Journal of Sport and Health Science* 4:198–207.
Parry-Jones, J. 2003. *Falconry: Care, Captive Breeding, and Conservation.* London: David and Charles PLC.
Price, A. 2002. *Raptors: The Eagles, Hawks, Falcons, and Owls of North America.* Lanham, Md.: Roberts Rinehart.

Sidebar: Striated Caracara

Meiburg, J.2004 The biogeography of Striated Caracaras Phalcoboenus australis. Master's thesis, University of Texas, Austin.

Sidebar: Antipredator Hunting

Gray Wolf. Defenders of Wildlife, http://www.defenders.org/wildlife _and_habitat/wildlife/wolves,_gray.php; accessed August 31, 2009.

Sidebar: Migration Theories

Alexander, R. M. 2002. The merits and implications of travel by swimming, flight, and running for animals of different sizes. *Integrative and Comparative Biology* 42:1060–1064.

Bauer, S., et al. 2009. Animal migration: Linking models and data beyond taxonomic limits. *Biology Letters* 5:4333–435.

Hedenström, A., and T. Alerstam. 1995. Optimal flight speeds of birds. *Philosophical Transactions of the Royal Society B: Biological Sciences* 348:471–487.

Houston, A. I. 1998. Models of optimal avian migration: State, time, and predation. *Journal of Avian Biology* 29:395–404.

Saino, N., and R. Ambrosini. 2008. Climatic connectivity between Africa and Europe may serve as a basis for phenotypic adjustment of migration schedules of trans-Saharan migratory birds. *Global Change Biology* 14:250–263.

Vrugt, J. A., et al. 2007. Pareto front analysis of flight time and energy use in long-distance bird migration. *Journal of Avian Biology* 38:432–442.

Sidebar: Sky Burial

Ash, N. 1990. *Flight of the Wind Horse: A Journey into Tibet.* London: Rider.

Kahn, S. 2006. An assessment of avian and other scavenging of an animal carcass at Katerniaghat Wildlife Sanctuary, District Bahraich, Uttar Pradesh, India, and its forensic implications. *Anil Aggrawal's Internet Journal of Forensic Medicine and Toxicology,* http://www.geradts .com/anil/ij/vol_007_no_001/others/thesis/roma.html; accessed September 2, 2009.

Logan, Pamela. Witness to a Tibetan sky-burial: A field report. California Institute of Technology Alumni Association. http://www .alumni.caltech.edu/~pamlogan/skybury.htm; accessed September 2, 2009.

Towering silence. Search, http://www.searchmagazine.org/Archives/ Back%20Issues/2008%20May-June/full-vultures.html; accessed September 2, 2009.

Yosef, R., and O. Bahat. 2000. Habitat loss and vultures: A case study from Israel. In *Raptors at Risk,* ed. R. D. Chancellor and B.-U. Meyburg. Blaine, Wash.: Hancock House.

Index

Accipiter cooperii. See Cooper's
Hawk
Accipiter gentilis. See Goshawk
Accipiter melanoleucus. See African
Black Sparrowhawk
Accipiter nisus. See European/Eur-
asian Sparrow Hawk
Accipiter striatus. See Sharp-shinned
Hawk
Aegolius acadicus. See Northern
Saw-whet Owl
Aegolius funereus. See Boreal Owl,
Tengmalm's Owl
Aegypius monachus. See Cinereous
Vulture
African Black Sparrowhawk,
138
African Crowned Hawk-Eagle, 30,
64, 71
African Fish Eagle, 24
African Harrier Hawk, 60
African Marsh Harrier, 73
African Pygmy Falcon, 20
Ahlquist, Jon, 12
Alexander, R. McNeill, 52
Alexander the Great, 124
Amadon, Dean, 27, 71, 79, 164
Ambrosini, Roberto, 53
American Kestrel, 13, 22, 25, 39,
71, 72, 90, *100*

American Museum of Natural
History, 8, 12, 27, 151, 164
American Ornithologists' Union,
12
American Sparrow Hawk, 10
Anas platyrhynchos. See Mallard
Andean Condor, 18, 26, 43, pl. B
Anna's Hummingbird, 50
Aplomado Falcon, 171
Aquila adalberti. See Spanish Impe-
rial Eagle
Aquila chrysaetos. See Golden Eagle
Aquila clanga. See Spotted Eagle
Aquila nipalensis. See Steppe Eagle
Arad, Zeev, 32
Archives of Falconry, 145
Arsenault, David, 73
Ash, Lydia, 127
Ash, Niema, 154
Asio flammeus. See Short-eared Owl
Athene cunicularia. See Burrowing
Owl
Audubon, John James, 9–10, 135,
140–153
Audubon, John Woodhouse, 150
Avian Protection Plan (AAP),
95–96

Bahat, Ofer, 108
Baird, Spencer Fullerton, 85

Bajazet, 130
Bald and Golden Eagle Protection
    Act, 172
Bald Eagle, 24, 36, *62*, 64, *67*, *96*,
    110–111, 158, 171–172
banding, 21, 57, 64, 82–83, 128,
    136–137, 169
bank voles, 164
Banks, Peter, 163–164
Barfield, Michael, 163
Barn Owl, 9, 16, 28, 37–39, 61, 65,
    78, 81–83, 162
Barnard, Capt. Charles, 166
Barred Owl, 102–103
Barrowclough, George, 12
Bartholomew, George A., 32
Barton, Nigel W. H., 28
beak, 1, 12, 16, 18, 24–25, 28, 36,
    58, 61, *62*, 64, 86, 92, 134
Bearded Vulture, 60, 154
Beebe, Frank, 76
Bendire, Major Charles E., 85
Berger, Cynthia, 27, 86
Berners, Dame Juliana, 126
Biblical Zoo, 108
Bildstein, Keith, 55, 57–58
bill, 15, 27, 36, *40*, 44, 60, 64, 87,
    pl. A
bird strike, 48–50
Birkhead, Tim, 69, 72
Black Kite, 25
Black Vulture, 19, 25, 43, 45
Black-browed Albatross, 165
Black-thighed Falconet, 20
Blakiston's Fish Owl, 17, *78*, 164
Blanco, Guillermo, 107
Boal, Clint, 91
Bobwhite Quail, 133
Bodio, Stephen, 122, 149, 159
Bond, Frank, 126–127, 144–145
bones, 6–8, 25, 28–30, 33, 51, 60,
    115, 123, 154, 157, 159, 171
Boreal Owl, 31, 78

Bowie, Rauri, 5
brain, 41, 44, 46
Branta canadensis. *See* Canada
    Geese
breeding, 3–4, 12, 23–24, 26, 35,
    52–54, 56, 59–61, *66*, 65–77,
    79–80, 83, 89–91, 101–103,
    105–106, 108, 113–116, 124,
    127–128, 132, 136, 144–145,
    158, 160–161, 165–167, 169
Briskie, James, 71
Broad-winged Hawk, 4
Bromham, Lindell, 13
Brown, Leslie, 27, 64, 71, 79, 164
brown hare, 137
Bubo bubo. *See* Eurasian Eagle Owl
Bubo (=Nyctea) scandiacus. *See*
    Snowy Owl
Bubo virginianus. *See* Great Horned
    Owl
Burrowing Owl, 31, 79, 101–102,
    pl. G
Buschhaus, Nancy, 71
Butler, Carol A., 58, *122*, *139*, 159,
    161
Buteo albonotatus. *See* Zone-tailed
    Hawk
Buteo buteo. *See* Common Buzzard
Buteo buteo vulpinus. *See* Steppe
    Buzzard
Buteo jamaicensis. *See* Red-tailed
    Hawk
Buteo jamaicensis harlani. *See*
    Harlan's Hawk
Buteo lagopus. *See* Rough-legged
    Hawk
Buteo lineatus. *See* Red-shouldered
    Hawk
Buteo platypterus. *See* Broad-winged
    Hawk
Buteo rufinus. *See* Long-legged
    Buzzard
Buteo solitarius. *See* Hawaiian Hawk

Buteo swainsoni. *See* Swainson's Hawk

Buteogallus meridionalis. *See* Savanna Hawk

Cade, Tom, 32, 51, 76, 122, 145
Calhim, Sara, 69
California Condor, 18, 87, 103, 171
Calypte anna. *See* Anna's Hummingbird
camouflage, 1
Canada Geese, 48
Canadian Peregrine Foundation, 79
canaries, 113
Canis lupus. *See* gray wolf
Capainolo, Peter, 25, 48, 77, 83, 99, 135, 141, 143, 159
Cape May Bird Observatory, 57
Caracara cheriway. *See* Crested Caracara
Carl VI, 130
Carroll, Karen "Kitty" Tolson, 128
cassowaries, 70
Castle, James, 101
Catesby, Mark, 147
Cathartes aura. *See* Turkey Vulture
Cathartes melambrotus. *See* Greater Yellow-headed Vulture
Catherine the Great, 126
Centre for Macroevolution and Macroecology (ANU), 13
Chapman, Frank M., 147–148
Chlamydotis undulate. *See* Houbara Bustard
Cinereous Vulture, 18, 107, 154
Circaetus. *See* Snake Eagle
Circus aeruginosus. *See* Marsh Harrier
Circus approximans. *See* Swamp Harrier
Circus cyaneus. *See* Northern Harrier

Circus ranivorus. *See* African Marsh Harrier
Clark, Howard, 102
Clethrionomys glareolus. *See* bank voles
cloaca, 29, 68, *69*, 69–72, 76–77
Colinus virginianus. *See* Bobwhite Quail
Collared Falconet, 20, pl. D
Collier, Julie, 93, 118
Columba livia. *See* Rock Pigeon
Columba palumbus. *See* Wood Pigeon
Common Buzzard, 4, 19, 28, 34
condors, 35, 103–104
Cooper's Hawk, *2*, 22, 62, 91, 98, 138, 148, 171
Coragyps atratus. *See* Black Vulture
Cortéz, Hernán, 130
cottontail rabbit, 133
Coturnix coturnix. *See* Coturnix Quail
Coturnix Quail, 118
Craighead, Frank, 60, 163
Craighead, John, 60, 163
Crane Hawk, 60, pl. D
Crested Caracara, 171, pl. D
Crested Pigeon, 91
crop, 26–29, 92, 106, 135, pl. A

Darwin, Charles, 13, 166
dichloro-diphenyl-trichloroethane (DDT), 76, 104–105, 160, 168
dinosaurs, 1, 5–8
diurnal raptors, 2–3, 16, 39, 42, 58, 61, 65, 72, 84, 101, 139, 167
diving, 15, 17, 43, 49, 50–51, 59, 64, 121, 139, 169
DNA, 9, 11–14, 26, 71, 73, 75
dogs, 5, 19, 47, 107, 131–132, 138, 140–141, 166
Drost, Rudolf, 57
Duck Hawk, 104

ducks, 35, 64, 70, 104, 129, 169
Duke, Gary, 93
Dwyer, James, 95

eagles, 1, 3, 9, 16, 18–27, 34–36,
    39, 41, 51, 59, 62–64, 68, 71,
    87, 95, 103, 108, 110–111, 114,
    120–121, 124, 127, 129–130,
    133, 144, 148, 156, 157, 172,
    pl. E
Eastern Screech Owl, 84, 161
Eberhard, William, 74
Edge, Rosalie Barrow, 148
eggs, 6, 12, 16, 60, 64, 66–68,
    70, 75–76, 78, 79, 83–86, 101,
    104–105, 108, 111, 114, 165,
    168
Egyptian Vulture, 60, 107
Eldegard, Katrine, 73
Elf Owl, 20
Elizabeth I, 126
Emlen, Steven, 58
Endangered Species Act, 102
Enderson, James, 145
Erritzoe, Johannes, 46
Eudyptes chrysocome. See Rockhop-
    per Penguin
Eurasian Black Vulture. *See* Cine-
    reous Vulture
Eurasian Eagle Owl, 18, 89
Eurasian Hobby, 4, 24, 28, 56
Eurasian Kestrel, 43
European Kestrel, 28
European rabbit, 137
European (Eurasian) Sparrow
    Hawk, 4, 13, 14, 22, 28, 47, 57,
    87, 89, 121, 150
European Starling, 23, 97
Evans, Humphrey Ap, 133
evolution, 1, 7–9, 12–14, 22, 43, 89,
    149, 154, 163; convergent evolu-
    tion, 1, 19
eyass, 76, 132, 143

Falco biarmicus. See Lanner Falcon
Falco cherrug. See Saker Falcon
Falco columbarius. See Merlin
Falco femoralis. See Aplomado
    Falcon
Falco islandicus. See Jer/Iceland
    Falcon
Falco mexicanus. See Prairie Falcon
Falco naumanni. See Lesser Kestrel
Falco novaeseelandiae. See New Zea-
    land Falcon
Falco peregrinus. See Peregrine
    Falcon
Falco peregrinus anatum. See Duck
    Hawk
Falco punctatus. See Mauritius
    Kestrel
Falco rusticolus. See Gyrfalcon
Falco sparverius. See American
    Kestrel; American Sparrow
    Hawk
Falco subbuteo. See Eurasian
    Hobby
Falco tinnunculus. See Eurasian
    Kestrel
Falco vespertinus. See Red-footed
    Falcon
Falcon Environment Services
    (FES), 143
falconer, 11, 14, 22, 26, 37, 44,
    49–51, *63*, 64, 75–76, 97,
    99, 118–120, 122, 124–132,
    135–144, 149, 159, 168–169;
    falconry, 47, 76, 94, 99, 113,
    119–133, 135–138, 142–146,
    153, 158
Falconet, 3, 23, pl. D
falcons: 1, 3–4, 9–10, 12, 14–16,
    20, 22, 24, 27, 29, *30*, 32, 34–35,
    37, 39, 42, 49–51, *51*, 59, 62, 64,
    70, 74–78, 87, 97–98, *100*, 104,
    118, 120–122, *123*, 124–125,
    127–133, 135–138, *139*,

140–141, *142*, 143, 148, *152–153*,
 156, 160, 165, 169, pl. D;
feathers, 5–8, 12, 15, 17, 28–33,
 35–39, 48, 57, 60, 82–84, 88,
 91–92, 94, 110–111, 113, 119,
 130, 134, 140–141, 148, 157,
 171–172; molting, 35–36, 119;
 plumage, 35–36, 82, 86, 92,
 150, 155, 171, preening, 36–37;
 rousing, 37
Field Museum of Natural History,
 5, 12
field voles, 164
finches, 113
fish, 1, 23–24, 37, 103
Fish Hawk, 24. *See also* Osprey
Fite, Katherine, 42
Flammulated Owl, 73, 80
flight, 8, 25
Flora and Fauna Act, 137
Ford, Emma, 128
Forest Falcon, 3, 39
Fox, Nick, 44
Franklin, Ken, 50
Frederick II of Hohenstaufen, 29,
 124, 131
Fuertes, Louis Agassiz, 131
Fyfe, Richard, 76, 145

Galeotti, Paolo, 171
Gallus gallus. See hen, domestic
Garamszegi, Lazlo, 46
geese, 35, 48–49, 70, 116
George, Jean Craighead, 128
Geranospiza caerulescens. See Crane
 Hawk
Giant Petrel, 165
Gill, Peter, 76
Glasier, Phillip, 133
Glaucidium californicum. See Pygmy
 Owl
Glaucidium gnoma. See Northern
 Pygmy Owl

Golden Eagle, 3, 19, 21, 25–27,
 39, 41, 47, 49, 95, 137–138,
 157–159, 169, 172
Gomez, Luis, 45
Gorney, Edna, 54
Goshawk, 4, 10, 14, 22, 62, 64, 73,
 89, 121, 129, 135, 137, 148
Gowaty, Patricia, 71
Grande, J. M., 72
gray wolf, 109
Great Grey Owl, 39, 80
Great Horned Owl, 12, 16, 25, 34,
 66, 80, 85, 87, 90, 148
Greater Yellow-headed Vulture,
 45
Griesser, Michael, 90
Griffiths, Carole, 12
Griffon Vulture, 32, 107–108
Groth, Jeff, 12, *33, 117*
grouse, 3, 27
Gymnogyps californianus. See Cali-
 fornia Condor
Gypaetus barbatus. See Bearded
 Vulture
Gypohierax angolensis. See Palm-nut
 Vulture
Gyps bengalensis. See Indian White-
 backed Vulture
Gyps fulvus. See Griffon Vulture
Gyps himalayensis. See Himalayan
 Griffon
Gyps indicus. See Indian Long-
 billed Vulture
Gyrfalcon, 3–4, 75–76, *122*, 125,
 130, 135, 151, *152–153*

habitat, 3–4, 24, 31, 43, 47, 52,
 61, 74, 76, 78–79, 90, 102–103,
 114, 124, 145, 162, 164–165,
 167–168, 172
Hackett, Shannon, 5, 12, 14
Haliaeetus albicilla. See White-tailed
 Sea Eagle

Haliaeetus leucocephalus. See Bald Eagle
Haliaeetus pelagicus. See Steller's Sea Eagle
Haliaeetus vocifer. See African Fish Eagle
Hamilton, James, 166–167
Hammerstrom, Fran, 128–129
Harlan, Richard, 9
Harlan's Hawk, 9
Harness, Rick, 95
Harpia harpyja. See Harpy Eagle
Harpy Eagle, 3, 18, 68
harriers, 4, 9–10, 16, 39, 89
Harris's Hawk, 47, *63,* 128, 135
Hawaiian Hawk, 171
Hawk Creek Wildlife Center, *122,* 157–158
Hawk Mountain Sanctuary, 55, 57, 156
hawks, 1, 9–11, 14–16, 18–20, 23, 25, 27–29, *30,* 38–39, 42–44, 47, 54, 57, 60, 62, 67, 79, 90–91, 98, 101, 111–112, 114, 119–121, 124–126, 128, 130, *131,* 133–135, 137–138, 140–144, 148, 156, 170, pl. A, pl. D, pl. H
HawkWatch International, 95
hearing, 1, 3, 38–39, 61, 101
Heintzelman, Donald, 155
hen, domestic, *70*
Henry II, 125
Heppenstall, John, 151
Himalayan Griffon, 154
Hingee, Mae, 91
Holt, Robert, 163
Honey Buzzard, 21, 56
horses, 85, 138
Houbara Bustard, 76
Houston, Alasdair, 52
Houston, David, 45
Hsu, Yu-Cheng, 73

hunting, 2–4, 14, 16–17, 19, 22–26, 32, 34, 38–39, 42, 44, 47, 55, 59–61, 72, 76, 79–80, 83–84, 89, 94, 96–99, 104, 109–110, 114, 116, 119–127, 129–133, 136–141, 143–145, 155, 159, 164, 169
husbandry, 113, 129, 137
hybrid, 75–76, *123,* 135–136

ibises, 12
imprinting, 77, 108, 116–118, 143
Indian Long-billed Vulture, 106
Indian White-backed Vulture, 106, 154
Indigo Bunting, 58
insects, 1, 3, 15, 19–20, 23–25, 30, 56, 67, 84, 97, 104, 107, 162, 164
International Association for Falconry and Conservation of Birds of Prey (IAF), 126–127, 144–145
International Birding and Research Center, 108
International Union for Conservation of Nature, 167
irruption, 24
Ito, Ryo, 89

Jer/Iceland Falcon, 151
John II, 130

Kanemitsu, 126
Keller, Hannah, *123,* 127
Kerlinger, Paul, 57
Ketupa blakistoni. See Blakiston's Fish Owl
Khan, Genghis, 125
Khan, Kublai, 125
King Cormorant, 166
King Vulture, 45, pl. B
Kirkwood, James K., 28
kites, 1, 4, 9, 120, pl. C, pl. F

kiwis, 70
Kochiku, 127
Korpimaki, Erkki, 43, 73

Laboratoire d'Ecologie, Ecole
    Normale Superieure, 20, 120
Lagopus. See Ptarmigan
Lagopus scoticus. See Red Grouse
Lanner Falcon, 4
Lanyu Scops Owl, 73
LeCroy, Mary, 151
Lemus, Jesus, 107
Lepus europaeus. See brown hare
Lessels, Catherine, 72
Lesser Kestrel, 24
Levy-Yamamori, Ran, 108
Linnaeus, Carl, 13, 147
lions, 61, 108
Logan, Pamela, 154
Lombardo, Michael, 71
Long-legged Buzzard, 80
Lorenz, Konrad, 116
Lorenz Center, 52
Los Alamos National Laboratory,
    52
Los Angeles Zoo, 103

Macdonald, Helen, 128
Macgillivray, William, 151
Macronectes halli. See Giant Petrel
Madagascar Paradise Flycatcher,
    90
Magrath, Robert, 91
Mallard, 137
mammals, 1, 3–4, 16, 20, 23–26,
    33–34, 41–42, 45–46, 68, 89,
    93, 101, 111, 162
Mannan, R. William, 95
Marsh Harrier, 56
Mary, Queen of Scots, 126
Mary of Canterbury, 126
mating, 4, 22, 61, 65–68, 70–74,
    79, 83, 116, 155, 158

Mauritius Kestrel, 3
McShea, William, 45
Megascops asio. See Eastern Screech
    Owl
Meiburg, Jonathan, 165
melanin, 6
Meng, Heinz, 76, 97, 145
Merlin, 4, 20, 22, 27, *33*, 97–99,
    100, 126–127, 135, 171
Mertz, Lisa, 12
metabolism, 23, 25–28, 31, 55, 104
mice, 15, 23, 25, 38–39, 43, 97,
    114, 118, 162, 170
Micrastur. See Forest Falcon
Micrathene whitneyi. See Elf Owl
Microhierax. See Falconet
Microhierax caerulescens. See Col-
    lared Falconet
Microhierax fringillarius. See Black-
    thighed Falconet
Microtus agrestis. See field voles
migration, 4, 10, 20, 24, 26, 32, 48,
    50, 52–59, 62, 71, 87, 99, 101,
    112, 124, 132, 138, 143, 148,
    153, 155–156, 160, 166, 168,
    170
Migratory Bird Treaty Act, 94, 101,
    136, 172
Miller, Yigal, 117
Milvus migrans. See Black Kite
Milvus milvus. See Red Kite
Mirza, Husam al-Dawlah Timur,
    142
Møller, Anders Pape, 46, 73
Montezuma, 130
Montgomerie, Robert, 71
Mori, Akira, 89
Morris, John, 75
Morrow, Jill, 128
Mosher, James, 32
Mougeot, Francois, 73
Museo Nacional de Ciencias Natu-
    rales, Madrid, 107

Museum of Natural History Vienna, 75
muskrats, 133

National Eagle Repository, 172
Natural History Museum, London, 38
Negro, J. J., 72
Neophron percnopterus. See Egyptian Vulture
nesting, 4, 16, 20, 30, 36, 60, 64–65, 68, 71–73, 78–89, 91–92, 94–96, 101, 103, 105–108, 110–111, 116, 118, 132, 143, 160–162, 164–165, 169, pl. C
New York State Ornithological Association, 160
New Zealand Falcon, 24
Nintoku, 126
Nittinger, Franziska, 75
nocturnal raptors, 2, 16, 39, 43–44, 46, 58, 60–61, 65, 72, 79, 101, 120, 139
North American Falconers' Association, 127, 131, 144
Northern Harrier, *4, 11*, 34, 39
Northern Hawk Owl, 2, 16
Northern Pygmy Owl, 20
Northern Saw-whet Owl, 78, 148, *161*
Northern Spotted Owl, 102, 171

Oceanites oceanicus. See Wilson's Storm Petrel
Ocyphaps lophotes. See Crested Pigeon
Oliphant, Lynn, 75
Olsen, Jerry, 23
Ondatra zibethicus. See muskrats
Oplurus cuvieri cuvieri. See spiny-tailed iguana
Oryctolagus cuniculus. See European rabbit

Osprey, 1, 9, 17–19, 21, 24, 37, 56, 59, *66,* 73, 103–105, 139, 156, 171
Ostrich, 29, 70
Otsuka, Noriko, 126
Otus elegans botelensis. See Lanyu Scops Owl
Otus flammeolus. See Flammulated Owl
Otus scops. See Scops Owl
owls, 1–2, 9, 12, 16–17, *17,* 22–23, 27–28, 31–32, 37–39, 41, 43–44, 60–61, 65–67, 78–86, 90, 92, 101–103, 114, 133, 139, 164, pl. G

Pale Male, 4
Palm-nut Vulture, 19, 23
Pandion haliaetus. See Osprey, Fish Hawk
Parabuteo unicinctus. See Harris's Hawk
Parks, Jim, 93, 118
parrots, 14, 113
Parry-Jones, Jemima, 60, 127–128
Passerina cyanea. See Indigo Bunting
passerines, 14, 52, 90
Patricelli, Gail, 90
Pauli, Amy, 93
Peabody Museum, 12
pellets, 23, 27–31, 83, 92
Peregrine Falcon, 3–4, 9, *10,* 15, 22–23, 26, 28, 35, 49–51, 53–54, *69,* 75–76, 79–80, 84, 89, 104–105, 121, 125, 135, 137, *139, 142,* 145, 151, 160–161, 168
Peregrine Fund, 145
Perisoreus infaustus. See Siberian Jay
Pernis apivorus. See Honey Buzzard
Peterson, Aaron, 69
Phalacrocorax albiventer. See King Cormorant

Phalcoboenus australis. See Striated Caracaras

Phasianus colchichus. See Ring-necked Pheasant

Phillip the Bold, 130

Pito, Ahizato, 124

Plumpton, David, 102

Polihierax. See Pygmy Falcon

Polihierax semitorquatus. See African Pygmy Falcon

Polo, Marco, 125

Polyboroides typus. See African Harrier Hawk

Pough, Richard, 148

prairie dogs, 101

Prairie Falcon, 75

Predatory Bird Group, 145

Price, Anne, 127–129

Procyon lotor. See raccoons

Protection of Wildlife Act, 137

Ptarmigan, 3

Pygmy Falcon, 3, 20

Pygmy Owl, 85, 87

rabbits, 23, 39, 47, 49, *63,* 133, 135, 137, 141

rabies, 93, 107

raccoons, 133, 169

Rachel, John, 110

rangle, 29

Raptor Education Foundation, 128

Raptor Watch, 55

red fox, 133

Red Grouse, 76

Red Kite, 28

Red-footed Falcon, 24, 54

Red-headed Vulture, 154

Red-shouldered Hawk, 80, 134, 171, pl. A

Red-tailed Hawk, 4, 9, 16, 19–20, 25–26, 34, 39, 47, 67, 134, 161, pl. A, pl. H

Redig, Patrick, 93

Rice, William, 39

Richardson, Malena, 73

Richter, Bernadette, 128

Ring-necked Pheasant, 133, 137

Robinson, Bill, 45

Robinson, Meg, 129

Rock Pigeon, 160

Rockhopper Penguin, 165

Rocky Mountain Arsenal National Wildlife Refuge, 172

Rosenfield-Wessels, Sheila, 42

Rostrhamus sociabilis. See Snail Kite

Rough-legged Hawk, 3, 59, 79

Sagittarius serpentarius. See Secretary Bird

Saino, Nicola, 53

Sakenokimi, 126

Saker Falcon, 4, 75–76, 96, *123,* 125, 143, *152*

Saladin, 124

San Diego Zoo, 103–104

Sarcogyps calvus. See Red-headed Vulture

Sarcoramphus papa. See King Vulture

Sattler, Helen Roney, 80

Savanna Hawk, 51

Save Our American Raptors (SOAR), 128

Schoultz, Elizabeth, 127–129

Scops Owl, 171

Scotopelia bouvieri. See Vermiculated Fishing Owl

Seaton, Richard, 24

Secretary Bird, 24, 27, 30–31, 34, 47, 59, pl. C

sexual dimorphism, reversed, 22

Shannon, L. M., 93

Sharp-shinned Hawk, 4, 26, 47, 148

Short-eared Owl, 3, 16, 79

Shukou, 127

Siberian Jay, 90
Sibley, Charles, 12
sight, 1, 38–39, 42, 170
Silvereye, 58
Simmons, Robert, 73
Slaght, Jonathan, *17*, *78*, 164
Snail Kite, 19, 24, 139, 171, pl. F
Snake Eagle, 24, 30
Snowy Owl, 3, 12, 16, 41, 79, 85, pl. G
Snyder, Helen, 21
Sonerud, Geir, 73
Spanish Imperial Eagle, 3
Spielberg, Steven, 7
spiny-tailed iguana, 90
Spotted Eagle, 25
Stager, Kenneth E., 45
Steller's Sea Eagle, *40*, pl. E
Stephanoaetus coronatus. See African Crowned Hawk-Eagle
Steppe (Eurasian) Buzzard, 23, 54, *55*
Steppe Eagles, 25
Stevens, Ronald, 75
Stewart, Debbie, 127
storks, 12
Strandberg, Roine, 56
Striated Caracara, 165–167
Strix nebulosa. See Great Grey Owl
Strix occidentalis caurina. See Northern Spotted Owl
Strix varia. See Barred Owl
Struthio camelus. See Ostrich
Strutt, James, 126
Sturnus vulgaris. See European Starling
Surmach, Sergei, 164
Surnia ulula. See Northern Hawk Owl
Sutton, George Miksch, 148
Swainson's Hawk (Grasshopper Hawk), 10, 19, 23, 54, *55*, 59, 111

Swamp Harrier, 28
swans, 70
Sylvilagus floridanus. See cottontail rabbit

Tachycineta bicolor. See Tree Swallow
taiga, 3–4
tails, 15–16, 18–21, 35–36, 68, *69*, 83, 98, 113, 134, 141, 165
talons, 1, 12, 16–17, 25, 27, 36–37, 50, *62*, 63–64, *67*, 68–70, 91, 156, 158, pl. A, C; claws, 1, 5, 38, 159
Tengmalm's Owl, 74
Terpsiphone mutate. See Madagascar Paradise Flycatcher
Thalassarche melanophrys. See Black-browed Albatross
Thiollay, Jean-Marc, 20, 120
Tieleman, Irene, 26
tiercel, 22
Timmons, Scott, 142
Tisch Family Zoo, Israel, 117
Tollan, Andrew, 28
Trapani, Josh, 30
Tree Swallow, 69
Trulio, Lynne, 101
Tucker, Vance, 42
Tunstall, John, 9
Turkey Vulture, *6*, 31–32, 34, 45, 92, 128
Turner, Alan, 8
Tyto alba. See Barn Owl

UN Educational, Scientific and Cultural Organization (UNESCO), 144–145
U.S. Fish and Wildlife Service, 60, 95, 103, 105, 114, 167

Vermiculated Fishing Owl, 17
veterinarian, 114, 129, 136, 156–157

Vinther, Jakob, 6
vision, 39, *40*, 40–44, 46, 50, 58, 101, 165
voles, 39, 43, 82, 164
Vrugt, Jasper A., 52
Vulpes vulpes. See red fox
Vultur gryphus. See Andean Condor
vultures, 1, 4, 9, 11–12, 18–19, 22–23, 25, 32, 34–35, 37, 43–45, 60, 79, 92, 106–108, *117*, 120, *125*, 139, 154, 156, pl. B

Waller, Renz, 76
Walsh, Stig, 38
Ward, Jennifer, 32
Weidensaul, Scott, 84–85, 148
Welty, Joel, 21
White, Clayton, 32, 80
White, Rev. Gilbert, 147
White-tailed Sea Eagle, 24
Widén, Per, 73
WildCare Incorporated, 114

Wilson, E. O., 149
Wilson's Storm Petrel, 166
Wiltschko, Wolfgang, 58
wings, 5, 8, 10–11, 14–18, 22, 25, 31–38, 65, 77, 83, 88, 91, 94, 97–98, *100*, 101–102, 113–115, 157–158, 171, pl. E
Wingspan Birds of Prey Trust, 127
wolves, 19, 107, 109–110
Wood Pigeon, 137
Woodbridge, Brian, 111
Woodford, James, 80
Wright, Mabel Osgood, 147
Wu Yi-Qun, 80

Yom-Tov, Yoram, 54
Yorzinski, Jessica, 90
Yosef, Reuven, 108

Zone-tailed Hawk, 85
Zosterops lateralis. See Silvereye

About the Authors

Peter Capainolo has had an interest in natural history, particularly ornithology, since boyhood. At age eighteen, he was granted one of the first falconry licenses issued by New York State. He studied zoology and practiced falconry under renowned ornithologist Heinz Meng at the State University of New York, College at New Paltz, and subsequently earned undergraduate and graduate degrees in biology. Currently he is senior scientific assistant in the Department of Ornithology at the American Museum of Natural History in New York City and an adjunct faculty member in the Department of Biology, the City College of the City University of New York. He is also a research associate and member of the board of trustees of the Long Island Natural History Museum and has been appointed to serve on the New York State Falconry Advisory Board.

Carol A. Butler is the originator and coauthor of the Rutgers University Press Animal Q&A series of natural history question-and-answer books. She is also the coauthor of *Salt Marshes: Natural and Unnatural History* (2009) and of *The Divorce Mediation Answer Book* (1999). Butler is a psychoanalyst and a mediator in private practice in New York City, an adjunct assistant professor at New York University in the Department of Applied Psychology, and a docent at the American Museum of Natural History. More information about her is available at www.seetheotherside.com and www.members.authorsguild.net/cabutler.